METHODS OF MICROARRAY DATA ANALYSIS IV

This page intentionally left blank

METHODS OF MICROARRAY DATA ANALYSIS IV

Edited by

JENNIFER S. SHOEMAKER
SIMON M. LIN
Duke Bioinformatics Shared Resource
Duke University Medical Center
Durham, NC, USA

Springer

eBook ISBN: 0-387-23077-7
Print ISBN: 1-4899-8784-3

Softcover re-print of the Hardcover 1st edition 2005

Boston

Created in the United States of America

Visit Springer's eBookstore at: http://ebooks.springerlink.com
and the Springer Global Website Online at: http://www.springeronline.com

Contents

Contributing Authors ix

Preface xiii

Acknowledgments xv

Introduction 1

CANCER: CLINICAL CHALLENGES AND OPPORTUNITIES 9
DAVID G. BEER

GENE EXPRESSION DATA AND SURVIVAL ANALYSIS 21
PETER J. PARK

THE NEEDED REPLICATES OF ARRAYS IN MICROARRAY EXPERIMENTS FOR RELIABLE STATISTICAL EVALUATION 35
SUE-JANE WANG, JAMES J. CHEN

POOLING INFORMATION ACROSS DIFFERENT STUDIES AND OLIGONUCLEOTIDE CHIP TYPES TO IDENTIFY PROGNOSTIC GENES FOR LUNG CANCER 51
JEFFREY S. MORRIS, GUOSHENG YIN, KEITH BAGGERLY, CHUNLEI WU, AND LI ZHANG

APPLICATION OF SURVIVAL AND META-ANALYSIS TO GENE EXPRESSION DATA COMBINED FROM TWO STUDIES 67
LINDA WARNOCK, RICHARD STEPHENS, JOANN COLEMAN

MAKING SENSE OF HUMAN LUNG CARCINOMAS GENE EXPRESSION DATA: INTEGRATION AND ANALYSIS OF TWO AFFYMETRIX PLATFORM EXPERIMENTS 81
XIWU LIN, DANIEL PARK, SERGIO ESLAVA, KWAN R. LEE, RAYMOND L.H. LAM, AND LEI A. ZHU

ENTROPY AND SURVIVAL-BASED WEIGHTS TO COMBINE AFFYMETRIX ARRAY TYPES AND ANALYZE DIFFERENTIAL EXPRESSION AND SURVIVAL 95
JIANHUA HU, GUOSHENG YIN, JEFFREY S. MORRIS, LI ZHANG, FRED A. WRIGHT

ASSOCIATING MICROARRAY DATA WITH A SURVIVAL ENDPOINT 109
SIN-HO JUNG, KOUROS OWZAR, STEPHEN GEORGE

DIFFERENTIAL CORRELATION DETECTS COMPLEX ASSOCIATIONS BETWEEN GENE EXPRESSION AND CLINICAL OUTCOMES IN LUNG ADENOCARCINOMAS 121
KERBY SHEDDEN AND JEREMY TAYLOR

PROBABILISTIC LUNG CANCER MODELS CONDITIONED ON GENE EXPRESSION MICROARRAY DATA 133
CRAIG FRIEDMAN, WENBO CAO, AND CHENG FAN

INTEGRATION OF MICROARRAY DATA FOR A COMPARATIVE STUDY OF CLASSIFIERS AND IDENTIFICATION OF MARKER GENES 147
DANIEL BERRAR, BRIAN STURGEON, IAN BRADBURY, C. STEPHEN DOWNES, AND WERNER DUBITZKY

USE OF MICROARRAY DATA VIA MODEL-BASED CLASSIFICATION IN THE STUDY AND PREDICTION OF SURVIVAL FROM LUNG CANCER 163
LIAT BEN-TOVIM JONES, SHU-KAY NG, CHRISTOPHE AMBROISE, KATRINA MONICO, NAZIM KHAN AND GEOFF MCLACHLAN

MICROARRAY DATA ANALYSIS OF SURVIVAL TIMES OF PATIENTS WITH LUNG ADENOCARCINOMAS USING ADC AND K-MEDIANS CLUSTERING 175
WENTING ZHOU, WEICHEN WU, NATHAN PALMER, EMILY MOWER, NOAH DANIELS, LENORE COWEN, AND ANSELM BLUMER

HIGHER DIMENSIONAL APPROACH FOR CLASSIFICATION OF LUNG CANCER MICROARRAY DATA 191
F. CRIMINS, R. DIMITRI, T. KLEIN, N. PALMER AND L. COWEN

MICROARRAY DATA ANALYSIS USING NEURAL NETWORK CLASSIFIERS AND GENE SELECTION METHODS 207
GAOLIN ZHENG, E. OLUSEGUN GEORGE, GIRI NARASIMHAN

A COMBINATORIAL APPROACH TO THE ANALYSIS OF DIFFERENTIAL GENE EXPRESSION DATA 223
MICHAEL A. LANGSTON, LAN LIN, XINXIA PENG, NICOLE E. BALDWIN, CHRISTOPHER T. SYMONS, BING ZHANG AND JAY R. SNODDY

GENES ASSOCIATED WITH PROGNOSIS IN ADENOCARCINOMA ACROSS STUDIES AT MULTIPLE INSTITUTIONS 239
ANDREW V. KOSSENKOV, GHISLAIN BIDAUT, AND MICHAEL F. OCHS

Index 255

This page intentionally left blank

Contributing Authors

Ambroise, Christophe, Laboratoire Heudiasyc, Centre National de la Recherche Scientifique, Compiegne, France

Baggerly, Keith, University of Texas, MD Anderson Cancer Center, Houston, TX

Baldwin, Nichole, Department of Computer Science, University of Tennessee, Knoxville, TN

Beer, David, General Thoracic Surgery, University of Michigan, Ann Arbor, MI

Berrar, Daniel, School of Biomedical Sciences, University of Ulster at Coleraine, Northern Ireland

Bidaut, Ghislain, Bioinformatics, Fox Chase Cancer Center, Philadelphia, PA

Blumer, Anselm, Computer Science, Tufts University, Medford, MA

Bradbury, Ian, School of Biomedical Sciences, University of Ulster at Coleraine, Northern Ireland

Cao, Wenbo, City University of New York

Chen, James, Division of Biometry and Risk Assessment, National Center for Toxicologic Research, U.S. Food and Drug Administration

Coleman, JoAnn, GlaxoSmithKline

Cowen, Lenore, Department of Computer Science, Tufts University, Medford, MA

Crimins, F., Department of Computer Science, Tufts University, Medford, MA

Daniels, Noah, Department of Computer Science, Tufts University, Medford, MA

Dimitri, R., Department of Computer Science, Tufts University, Medford, MA

Downes, C. Stephen, School of Biomedical Sciences, University of Ulster at Coleraine, Northern Ireland

Dubitzky, Werner, School of Biomedical Sciences, University of Ulster at Coleraine, Northern Ireland

Eslava, Sergio, Biomedical Data Sciences, GlaxoSmithKline, Collegeville, PA

Fan, Cheng, UNC-Chapel Hill, Chapel Hill, NC

Friedman, Craig, NYU Courant Institute of Mathematical Sciences

George, E.O., Mathematical Sciences Department, University of Memphis, Memphis, TN

George, Stephen, Department of Biostatistics and Bioinformatics, Duke University Medical Center, Durham, NC

Hu, Jianhua, UNC-Chapel Hill, Chapel Hill, NC

Jones, Liat, Department of Mathematics and Institute for Molecular Bioscience, Univeristy of Queensland, Brisbane, Australia

Jung, Sin-Ho, Department of Biostatistics and Bioinformatics, Duke University Medical Center, Durham, NC

Khan, Nazim, Department of Mathematics, Univeristy of Queensland, Brisbane, Australia

Klein, T., Department of Computer Science, Tufts University, Medford, MA

Kossenkov, Andrew, Bioinformatics, Fox Chase Cancer Center, Philadelphia, PA

Lam, Raymond, Biomedical Data Sciences, GlaxoSmithKline, Collegeville, PA

Langston, Michael, Department of Computer Science, University of Tennessee, Knoxville, TN

Lee, Kwan, Biomedical Data Sciences, GlaxoSmithKline, Collegeville, PA

Lin, Lan, Department of Computer Science, University of Tennessee, Knoxville, TN

Lin, Xiwu, Biomedical Data Sciences, GlaxoSmithKline, Collegeville, PA

McLachlan, Geoff, Department of Mathematics and Institute for Molecular Bioscience, Univeristy of Queensland, Brisbane, Australia

Monico, Katrina, Department of Mathematics and Institute for Molecular Bioscience, Univeristy of Queensland, Brisbane, Australia

Morris, Jeffrey, University of Texas, MD Anderson Cancer Center, Houston, TX

Mower, Emily, Computer Science, Tufts University, Medford, MA

Narasimhan, G., School of Computer Science, Florida International University, Miami, FL

Ng, Shu-Kay, Department of Mathematics, Univeristy of Queensland, Brisbane, Australia

Ochs, Michael, Bioinformatics, Fox Chase Cancer Center, Philadelphia, PA

Owzar, Kouros, Department of Biostatistics and Bioinformatics, Duke University Medical Center, Durham, NC

Palmer, Nathan, Department of Computer Science, Tufts University, Medford, MA

Park, Daniel, Biomedical Data Sciences, GlaxoSmithKline, Collegeville, PA

Park, Peter, Children's Hospital Informatics Program and Harvard-Partners Center for Genetics and Genomics, Boston, MA

Peng, Xinxia, Graduate School of Genome Science and Technology, University of Tennessee, Knoxville, TN

Shedden, Kerby, Department of Statistics, University of Michigan, Ann Arbor, MI

Snoddy, Jay, Life Sciences Division, Oak RidgeNational Laboratory, Oak Ridge, TN

Stephens, Richard, GlaxoSmithKline

Sturgeon, Brian, School of Biomedical Sciences, University of Ulster at Coleraine, Northern Ireland

Symons, Christopher, Department of Computer Science, University of Tennessee, Knoxville, TN

Taylor, Jeremy, Department of Biostatistics, University of Michigan, Ann Arbor, MI

Wang, Sue-Jane, Division of Biometrics II, Office of Biostatistics, Office of Pharmacoepidemiology and Statistical Science, Center for Drug Evaluation and Research, U.S. Food and Drug Administration

Warnock, Linda, GlaxoSmithKline

Wright, Fred, UNC-Chapel Hill, Chapel Hill, NC

Wu, Chunlei, University of Texas, MD Anderson Cancer Center, Houston, TX

Wu, Weichen, Department of Computer Science, Tufts University, Medford, MA

Yin, Guosheng, University of Texas, MD Anderson Cancer Center, Houston, TX

Zhang, Bing, Life Sciences Division, Oak RidgeNational Laboratory, Oak Ridge, TN

Zhang, Li, University of Texas, MD Anderson Cancer Center, Houston, TX

Zheng, Gaolin., School of Computer Science, Florida International University, Miami, FL

Zhou, Wenting, Department of Computer Science, Tufts University, Medford, MA

Zhu, Lei, Biomedical Data Sciences, GlaxoSmithKline, Collegeville, PA

Preface

The fourth CAMDA conference, held in November 2003, focused on lung cancer data sets with a survival endpoint. In this volume, we highlight three tutorial papers to assist with a basic understanding of lung cancer, a review of survival analysis in the gene expression literature, and a paper on replication. In addition, 14 papers presented at the conference are included in this volume. Each paper was peer-reviewed and returned to the author for further revision. As editors, we have provided comments to the authors to encourage clarity and expansion of ideas.

As always, we do not propose these methods as the *de facto* standard for analysis of microarray data. Like the conference, they provide a point for continued discussion in an evolving field. Please join us for a future CAMDA conference to add your ideas to this discussion.

Jennifer S. Shoemaker

Simon M. Lin

This page intentionally left blank

Acknowledgments

The editors thank the contributing authors for their very fine work. We also acknowledge Emily Allred for her unflagging diligence and assistance in bringing this volume together, as well as for her hard work and dedication in organizing the CAMDA conference. We thank our supporters at Duke University: The Duke Comprehensive Cancer Center and the Center for Bioinformatics and Computational Biology. The CAMDA conference would not be possible without the contributions of the scientific committee and other reviewers (listed below) who contribute to the scientific review process. Our thanks for the time they commit to CAMDA. We especially thank our corporate sponsors for the generous support: North Carolina Biotechnology Center, GlaxoSmithKline and The Scientist. We gratefully acknowledge the North Carolina Biotechnology Center for providing a generous meeting grant. Finally, we offer a very large thanks to Kim Johnson, who has organized all four of the CAMDA conferences, and who co-edited, with Simon Lin, the previous three volumes. She has been very generous with her time and input to help this volume come to fruition.

Reviewers

Andrew Allen (Duke)
Bruce Aranow (U Cincinnati)
Chris Basten (NCSU)
Georgiy Bobashev (RTI)
Philippe Broett (INSERM)
Yong Chen (Wake Forest)
Kevin Coombes (MDACC)

Chris Corton (ToxicoGenomics)
Robert Delongchamp (NCTR)
Joaquin Dopazo (CNIO)
J. Gormley (MBI)
Greg Grant (U Penn)
Susan Halabi (Duke)
Wendell Jones (Expression Analysis,Inc.)
Michael Kelley (Duke)
Elana Kleymenova (CIIT)
Geoff McLachlan (U Queensland)
Dahlia Nielsen (NCSU)
Michael Ochs (Fox Chase)
Markus Ringner (Lund University)
Thomas Wu (Genentech)
Dmitri Zaykin (GlaxoSmithKline)

INTRODUCTION

After years of development since late nineties, microarray has become an established platform for high-throughput query of the transcriptome. The Critical Assessment of Microarray Data Analysis (CAMDA) conference continues serving as an annual forum for practitioners to exchange ideas and compare notes [Wigle et al., 2004]. The fourth CAMDA was held in November, 2003 with 145 researchers from 11 countries in attendance. As always, we were amazed by the new insights gained from reanalyzing published data sets. Following the CAMDA tradition, all papers analyzed the same designated data set, and the Best Presentation was voted by attendees and Scientific Committee members at the end of the conference. The CAMDA'03 Best Presentation went to:

> Jeffrey S. Morris, Guosheng Yin, Keith Baggerly, Chulei Wu, and Li Zhang, from M.D. Anderson Cancer Center for their paper "Pooling Information Across Different Studies and Oligonucleotide Chip Types to Identify Prognostic Genes for Lung Cancer".

With the help of the Scientific Committee, we compiled 14 papers from CAMDA'03 into this volume. In addition to research papers, we added three tutorials for beginners. In Chapter 1, Dr. David Beer (University of Michigan) summarized the research goals and challenges of lung cancer. Dr. Peter Park (Harvard) discussed the basic ideas and practical difficulties of associating microarray data with survival data in Chapter 2. Drs. Sue-Jane Wang and James Chen (FDA) discussed the importance of replicates in experiment design for reliable statistical evaluation in Chapter 3.

CAMDA 03 DATA SET

Lung cancer is the leading cause of cancer death worldwide [Jemal et al., 2003]. In contrast to survival improvements of breast cancer and prostate cancer in the past three decade, the five-year survival rate of lung cancer remained below 15% [Borczuk et al., 2003]. Experiments utilizing microarrays are expected to contribute to discoveries of biological mechanisms of lung cancer that might contribute to its poor outcome. The scientific committee of CAMDA selected four representative data sets published between 2001 and 2002 (Table 1) as the challenge data set for CAMDA'03.

Table 1. Microarray profiling of lung cancers.

	Paper Title	Author and Year
CAMDA 03 data sets		
"Harvard"	Classification of human lung carcinomas by mRNA expression profiling reveals distinct adenocarcinoma subclasses.	Bhattacharjee et al., 2001
"Stanford"	Diversity of gene expression in adenocarcinoma of the lung	Garber et al., 2001
"Michigan"	Gene-expression profiles predict survival of patients with lung adenocarcinoma	Beer et al., 2002
"Ontario"	Molecular profiling of non-small cell lung cancer and correlation with disease-free survival	Wigle et al., 2002
Other recent studies		
	Non-small-cell lung cancer molecular signatures recapitulate lung developmental pathways	Borczuk et al., 2003
	Gene expression profiling of normal human pulmonary fibroblasts following coculture with non-small-cell lung cancer cells reveals alterations related to matrix degradation, angiogenesis, cell growth and survival	Fromigue et al., 2003
	Expression profiles of non-small cell lung cancers on cDNA microarrays: identification of genes for prediction of lymph-node metastasis and sensitivity to anti-cancer drugs	Kikuchi et al., 2003
	cDNA microarray analysis of gene expression in pathologic Stage IA nonsmall cell lung carcinomas	Nakamura et al., 2003
	Transcriptional gene expression profiling of small cell lung cancer cells	Pedersen et al., 2003
	pRB2/p130 target genes in non-small lung cancer cells identified by microarray analysis	Russo et al., 2003
	Differentially expressed apoptotic genes in early stage lung adenocarcinoma predicted by expression profiling	Singhal et al., 2003
	MMP expression profiling in recurred stage IB lung cancer	Cho et al., 2004

From the recent studies in Table 1, we can see the research trend shifting from proof-of-concept study of the technology platform to more detailed investigation of lung cancer biology and its therapeutic implications. In addition to the continued investigation of lung cancer by transcriptional profiling, lung cancer has been studied with other high-throughput techniques, such as proteomics [Chen et al., 2003; Howard et al., 2003], loss of heterozygosity test [Massion et al., 2002; Janne et al., 2004] and tissue microarrays [Sugita et al., 2002; Haedicke et al., 2003]. All these advances in experiment biology continue to challenge bioinformatics in terms of data volume, complexity and integration.

FROM CLASSIFICATION TO SURVIVAL MODELING

In the previous years of CAMDA, discussions focused on classification [Lin and Johnson, 2002], pattern extraction [Lin and Johnson, 2002], and data quality control [Johnson and Lin, 2003]. CAMDA'03 initiated a new line of investigation of modeling survival data. Survival data adds complexity to the already complicated data analysis problem with censoring (see more discussion in tutorial Chapter 2). A comparison of this new task with previous tasks is summarized in Table 2. The goal of biomarker discovery is to find a small number of genes for developing rapid and low-cost clinical tests, whereas the goal of prediction is for diagnostics and prognostics.

Table 2. Tasks of classification and survival modeling.

	Biomarker discovery	Prediction
Classification	Find genes associated with a class of tumor	Predict the class given an expression profile
Survival Modeling	Find genes associated with survival time	Predict the patient survival given an expression profile

In Chapter 4, Morris et al. noticed an increased statistical power when combining data from different studies into a larger cohort. Warnock et al. (Chapter 5) proposed a meta-analysis approach to aggregate p-values from different studies. To associate gene expression with survival, investigators used different methods, including the Cox proportional hazard model [Tableman and Kim, 2004]. Lin et al. used survival tree in Chapter 6; Hu et al. proposed a weighted t-test to take both the cancer v.s. normal difference and cancer survival into consideration (Chapter 7); and Jung et al. (Chapter 8) used a nonparametric measure between a continuous variable and a survival variable and discussed the control of family-wise error rate in multiple testing. The paper of Jung et al. established a foundation for future studies of sample size estimation. This experimental design issue is of particular interest to clinical researchers who are planning to use microarrays in clinical trails. In Chapter 9, Shedden et al. associated clinical outcome with a novel differential correlation measure; and Friedman et al. approached the problem with developed tools and theories in econometrics (Chapter 10).

Berrar et al. (Chapter 11) and Zheng et al. (Chapter 15) investigate machine learning techniques to predict survival risk groups. Jones et al. applied model-based clustering to microarray data and demonstrated the association between the patient cluster and survival time (Chapter 12). In Chapter 13, Zhou et al. studied k-medians and approximate distance clustering in the context of survival analysis. In Chapter 14, the same group investigated the cancer-type classification problem. Zheng et al. further studied gene selection problem and compared the performance of different classifiers in Chapter 15. Langston et al. modeled the patient and gene relationship using edge-weighted graphs (Chapter 16). Kossenkov et al. used Bayesian Decomposition to find expression signatures related to patient prognosis in Chapter 17.

FROM DATA MODELING TO KNOWLEDGE MODELING

Discussion of biological relvevance is a key component of the CAMDA challenge. Previous reports in the literature of tumor biology were extensively used as supports for statistical findings. In addition, gene ontology (Chapter 5, 15, 16, and 17), GeneAtlas and OMIM database (Chapter 11) were used to assess the biological relevance.

To model biological knowledge formally in terms of signaling pathways, protein-protein interactions, and protein-drug interactions has been on the new roadmap of NIH [Zerhouni, 2003]. A recent study from Creighton et al. [2003] successfully utilized Locuslink and KEGG to suggest the loss of differentiation as a mechanism in lung adenocarcinomas. However, public resources, such as the LocusLink and KEGG, are still lacking in terms of quality, structure, and coverage for primetime use of knowledge modeling. Commercial entities are trying to fill in the gap by providing hand-annotated databases with higher quality, well defined ontology structure, and broader coverage. Such systems include Ingenuity Pathways Knowledge Base from Ingenuity (Mountain View, CA) and PathArt database from Jubilant Biosys (Columbia, MD). PathArt has been integrated into the PathwayAssist product from Ariadne (Rockville, MD) and the microarray analysis package from SpotFire (Somerville, MA). These integration tools provide a convenient way for end-users to model the functional pathways.

With the aid of these tools and many ongoing projects to develop similar tools, we will expect to see more efforts in modeling signaling pathways in the near future.

GRID-ENABLED SCIENCE

To address new research challenges in the post-genomic era, the National Cancer Institute launched the Cancer Biomedical Informatics Grid (caBIG) project. Dr. Kenneth Beutow from NCI introduced this project in the keynote address of CAMDA'03. CaBIG is an informatics platform to connect organizations and individuals by a data and computational grid [Nature News, 2004]. Many attendees were excited by its design principle of open access and common standards in bioinformatics. We are expecting to see the caBIG project nurturing transdisciplanry team science by sharing genomics data, tools, and computational infrastructure.

The caBIG project, together with the e-Science program of the United Kingdom, reflects the need from the scientific community to have access to large data sets and high-performance computing resources by grid computing [Butler, 2000]. We shall see early fruits from these projects in the future CAMDAs to come.

SUMMARY

Bioinformatics plays a key role in analyzing genomics data to fight life-threatening diseases. Our next CAMDA conference will focus on microarray studies of malaria. According to data from the Center of Disease Control of year 2000, malaria put 40% people in the world at risk; 300-500 million cases of infection were estimated yearly. We will see how microarrays help contribute to solving this public health problem, and how bioinformatics can help make sense of the data.

WEB COMPANION

Additional information can be found at the CAMDA website.

http://camda.duke.edu

You can try out algorithms in this book by following the download link from the authors; or, you can download the CAMDA'03 benchmark data set and run you new algorithms against it. In addition, conference slide presentations with color version of several figures can be found. Please also check the website for call for papers and announcements about the next conference.

REFERENCES

Beer, D. G., S. L. Kardia, C. C. Huang, T. J. Giordano, A. M. Levin, D. E. Misek, L. Lin, G. Chen, T. G. Gharib, D. G. Thomas, M. L. Lizyness, R. Kuick, S. Hayasaka, J. M. Taylor, M. D. Iannettoni, M. B. Orringer and S. Hanash (2002). Gene-expression profiles predict survival of patients with lung adenocarcinoma. Nat Med 8(8): 816-24.

Bhattacharjee, A., W. G. Richards, J. Staunton, C. Li, S. Monti, P. Vasa, C. Ladd, J. Beheshti, R. Bueno, M. Gillette, M. Loda, G. Weber, E. J. Mark, E. S. Lander, W. Wong, B. E. Johnson, T. R. Golub, D. J. Sugarbaker and M. Meyerson (2001). Classification of human lung carcinomas by mRNA expression profiling reveals distinct adenocarcinoma subclasses. Proc Natl Acad Sci U S A 98(24): 13790-5.

Borczuk, A. C., L. Gorenstein, K. L. Walter, A. A. Assaad, L. Wang and C. A. Powell (2003). Non-small-cell lung cancer molecular signatures recapitulate lung developmental pathways. Am J Pathol 163(5): 1949-60.

Butler, D. (2000). Europe joins race to turn the Internet info one vast computer. Nature 403: 213.

Chen, G., T. G. Gharib, H. Wang, C. C. Huang, R. Kuick, D. G. Thomas, K. A. Shedden, D. E. Misek, J. M. Taylor, T. J. Giordano, S. L. Kardia, M. D. Iannettoni, J. Yee, P. J. Hogg, M. B. Orringer, S. M. Hanash and D. G. Beer (2003). Protein profiles associated with survival in lung adenocarcinoma. Proc Natl Acad Sci U S A 100(23): 13537-42.

Cho, N. H., K. P. Hong, S. H. Hong, S. Kang, K. Y. Chung and S. H. Cho (2004). MMP expression profiling in recurred stage IB lung cancer. Oncogene 23(3): 845-51.

Fromigue, O., K. Louis, M. Dayem, J. Milanini, G. Pages, S. Tartare-Deckert, G. Ponzio, P. Hofman, P. Barbry, P. Auberger and B. Mari (2003). Gene expression profiling of normal human pulmonary fibroblasts following coculture with non-small-cell lung cancer cells reveals alterations related to matrix degradation, angiogenesis, cell growth and survival. Oncogene 22(52): 8487-97.

Garber, M. E., O. G. Troyanskaya, K. Schluens, S. Petersen, Z. Thaesler, M. Pacyna-Gengelbach, M. van de Rijn, G. D. Rosen, C. M. Perou, R. I. Whyte, R. B. Altman, P. O. Brown, D. Botstein and I. Petersen (2001). Diversity of gene expression in adenocarcinoma of the lung. Proc Natl Acad Sci U S A 98(24): 13784-9.

Haedicke, W., H. H. Popper, C. R. Buck and K. Zatloukal (2003). Automated evaluation and normalization of immunohistochemistry on tissue microarrays with a DNA microarray scanner. Biotechniques 35(1): 164-8.

Howard, B. A., M. Z. Wang, M. J. Campa, C. Corro, M. C. Fitzgerald and E. F. Patz, Jr. (2003). Identification and validation of a potential lung cancer serum biomarker detected by matrix-assisted laser desorption/ionization-time of flight spectra analysis. Proteomics 3(9): 1720-4.

Janne, P. A., C. Li, X. Zhao, L. Girard, T. H. Chen, J. Minna, D. C. Christiani, B. E. Johnson and M. Meyerson (2004). High-resolution single-nucleotide polymorphism array and clustering analysis of loss of heterozygosity in human lung cancer cell lines. Oncogene 23(15): 2716-26.

Jemal, A., T. Murray, A. Samuels, A. Ghafoor, E. Ward and M. J. Thun (2003). Cancer statistics, 2003. CA Cancer J Clin 53(1): 5-26.

Johnson, K. F. and S. M. Lin (2003). Methods of microarray data analysis III : papers from CAMDA '02. Boston, Mass., Kluwer Academic Publishers;.

Kikuchi, T., Y. Daigo, T. Katagiri, T. Tsunoda, K. Okada, S. Kakiuchi, H. Zembutsu, Y. Furukawa, M. Kawamura, K. Kobayashi, K. Imai and Y. Nakamura (2003). Expression profiles of non-small cell lung cancers on cDNA microarrays: identification of genes for prediction of lymph-node metastasis and sensitivity to anti-cancer drugs. Oncogene 22(14): 2192-205.

Lin, S. M. and K. F. Johnson (2002). Methods of microarray data analysis : papers from CAMDA '00. Boston, Kluwer Academic Publishers.

Lin, S. M. and K. F. Johnson (2002). Methods of microarray data analysis II : papers from CAMDA '01. Boston, Kluwer Academic Publishers.

Massion, P. P., W. L. Kuo, D. Stokoe, A. B. Olshen, P. A. Treseler, K. Chin, C. Chen, D. Polikoff, A. N. Jain, D. Pinkel, D. G. Albertson, D. M. Jablons and J. W. Gray (2002). Genomic copy number analysis of non-small cell lung cancer using array comparative genomic hybridization: implications of the phosphatidylinositol 3-kinase pathway. Cancer Res 62(13): 3636-40.

Nakamura, H., H. Saji, A. Ogata, M. Hosaka, M. Hagiwara, T. Saijo, N. Kawasaki and H. Kato (2003). cDNA microarray analysis of gene expression in pathologic Stage IA nonsmall cell lung carcinomas. Cancer 97(11): 2798-805.

Nature News (2004). Making data dreams come true. Nature 428: 239.

Pedersen, N., S. Mortensen, S. B. Sorensen, M. W. Pedersen, K. Rieneck, L. F. Bovin and H. S. Poulsen (2003). Transcriptional gene expression profiling of small cell lung cancer cells. Cancer Res 63(8): 1943-53.

Russo, G., P. P. Claudio, Y. Fu, P. Stiegler, Z. Yu, M. Macaluso and A. Giordano (2003). pRB2/p130 target genes in non-small lung cancer cells identified by microarray analysis. Oncogene 22(44): 6959-69.

Singhal, S., K. M. Amin, R. Kruklitis, M. B. Marshall, J. C. Kucharczuk, P. DeLong, L. A. Litzky, L. R. Kaiser and S. M. Albelda (2003). Differentially expressed apoptotic genes in early stage lung adenocarcinoma predicted by expression profiling. Cancer Biol Ther 2(5): 566-71.

Sugita, M., M. Geraci, B. Gao, R. L. Powell, F. R. Hirsch, G. Johnson, R. Lapadat, E. Gabrielson, R. Bremnes, P. A. Bunn and W. A. Franklin (2002). Combined use of oligonucleotide and tissue microarrays identifies cancer/testis antigens as biomarkers in lung carcinoma. Cancer Res 62(14): 3971-9.

Tableman, M. and J. S. Kim (2004). Survival analysis using S : analysis of time-to-event data. Boca Raton, Fla., Chapman & Hall/CRC.

Wigle, D. A., I. Jurisica, N. Radulovich, M. Pintilie, J. Rossant, N. Liu, C. Lu, J. Woodgett, I. Seiden, M. Johnston, S. Keshavjee, G. Darling, T. Winton, B. J. Breitkreutz, P. Jorgenson, M. Tyers, F. A. Shepherd and M. S. Tsao (2002). Molecular profiling of non-small cell lung cancer and correlation with disease-free survival. Cancer Res 62(11): 3005-8.

Wigle, D. A., M. Tsao and I. Jurisica (2004). Making sense of lung-cancer gene-expression profiles. Genome Biol 5(2): 309.

Zerhouni, E. (2003). Medicine. The NIH Roadmap. Science 302(5642): 63-72.

Chapter 1

CANCER: CLINICAL CHALLENGES AND OPPORTUNITIES

Lung Cancer Gene Profiling

David G. Beer
General Thoracic Surgery. University of Michigan, Ann Arbor MI 48109

Abstract: Lung cancer is the leading cause of cancer death. Gene expression profiling of lung cancer may provide one method to increase our understanding of this very heterogeneous disease and potentially identify new approaches for early diagnosis, prognosis or treatment. This brief review examines some of the issues that are associated with the clinical presentation of this disease and some important questions that might be addressed using gene expression profiling experiments when applied to human lung cancer.

Key words: Lung cancer; non-small cell lung cancer; mRNA; gene expression; gene profiles

1. INTRODUCTION

Gene expression profiling allows the examination of thousands of genes in a cell or tumor sample. The analysis of the patterns of gene expression and the identification of specific genes and pathways has the potential to help uncover biologically meaningful information. Recently published studies have utilized these technologies and applied these tools to lung cancer [Petersen et al., 2000; Garber et al., 2001; Bhattacharjee et al., 2001; Nacht et al., 2001; Beer et al., 2002; McDoniels-Silvers et al., 2002; Heighway et al., 2002; Miuri et al., 2002; Wigle et al., 2002; Wikman et al., 2002; Nakamura et al., 2003; Kikuchi et al., 2003; Difilippantonio et al., 2003; Yamataga et al., 2003; Borczuk et al., 2003]. Many of these analyses have revealed gene expression patterns that correlate with known histological patterns as well as reveal potential subgroups among lung adenocarcinomas that differ based on patient outcome. The ability to identify gene expression patterns or profiles that correlate with the biological aggressiveness of lung adenocarcinomas [Garber et al., 2001; Bhattacharjee et al., 2001; Beer et al., 2002; Wigle et al., 2002] suggests that these approaches may allow extraction of clinically useful information beyond that provided by a pathological assessment of the cancers alone.

Given the tremendous interest in the research community and the great potential for gene expression analyses to uncover new information about lung cancer, it is important to identify the clinical questions that are relevant to this disease and where these experimental approaches may prove useful in addressing these challenging questions. In this brief review some of these questions will be discussed with the goal that analytical approaches might be developed that focus on these biologically and clinically relevant problems in lung cancer.

1.1 Lung Cancer: A World Wide Problem

Lung cancer is the leading cause of cancer deaths in both men and women in the U.S. with nearly 160,000 cases per year [Jemal et al., 2003]. Importantly, this cancer is also one of the leading types worldwide accounting for 921,000 or 17.8% of all cancer deaths and greater than the mortality associated with breast, prostate, colon and pancreatic cancer combined [Ferlay et al., 2001]. There is a close relationship between lung cancer incidence (new cases/year) and mortality (deaths/year) due to the poor overall 5-year median survival, which is approximately 15% in the U.S. and only 8% in Europe and developing countries. As has been extensively demonstrated, the main cause of lung cancer is the consumption of tobacco products especially cigarette smoking. Of particular concern is that the rate of active smoking is increasing among high school students [US DHHS, 1998]. Since there is a long latency period between carcinogen exposure and cancer development, even if the total removal of all cigarettes were possible today there would still be an epidemic of lung cancer for many decades.

The majority of lung cancers occur in the upper lobes of the lung and most often present at a relatively advanced stage at first diagnosis. Stage I tumors account for approximately 13%, stage II for 10%, stage III for 44% and stage IV for 32% of all new lung cancers [Bulzebruck et al., 1992]. Thus most of these tumors when first seen by a physician have already metastasized either within the lung to associated lymph nodes (stages II and III) or to distant sites (stage IV). Given that the 5-year survival rates of patients dramatically decreases with increasing stage, the best hope for increasing patient survival is the early detection of lung cancer. This is one area that gene expression profiling may play a major role. Genes that are highly expressed in lung cancers, or that may be specific to each of the various types may be useful in defining new markers for monitoring patients at greatest risk and identifying cancer early when it is most effectively treated.

1.2 Multiple Types of Lung Cancer

Lung cancer is an extremely heterogeneous disease potentially more so than other cancer types. The World Health Organization histological classification divides lung cancer into two main types, small cell and non-small cell lung cancer [Brambilla et al., 2001]. The small cell type is characterized by small cells with a high nuclear to cytoplasmic ratio and the expression of a number of neuroendocrine gene products. These cancers can either be of the pure or mixed varieties. The non-small cell lung cancers account for the majority of lung cancer and include large cell, squamous cell, adenocarcinoma, adenocystic and carcinoids as the major subtypes [Travis et al., 1995]. Within large cell tumors there are both giant and clear cell types, within the squamous cancers there are epidermoid and the spindle cell types, and among adenocarcinomas there are the acinar, papillary, mucinous and bronchioloalveolar types. Often there may be tumors that display mixtures of these histological patterns. The lung is comprised of a large number of different cell types. The varying types and subtypes of lung cancer likely reflect these varied cell origins as well as histological and morphological features that may change during the loss of differentiated functions and genomic alterations associated with tumor growth and progression.

Although the histological features used to pathologically classify lung cancers have allowed some understanding of this disease, morphological features alone are insufficient to define the behavior of these tumors. The survival of patients with small cell, large cell, squamous cell and adenocarcinoma is very poor and unfortunately, these are the lung cancer types with the highest age-adjusted incidences [Zheng et al., 1994]. Analyses of gene expression profiles that have included the different histological types of lung cancers revealed gene expression patterns that appear to recapitulate these main pathological types [Garber et al., 2001; Bhattacharjee et al., 2001]. This indicates that as expected, the different lung cancer types have distinct genes that can distinguish them from the other lung cancers. Importantly, among the adenocarcinomas, several subgroups were also identified that appear to differ not only in gene expression but also as related to the patients overall survival [Garber et al., 2001; Bhattacharjee et al., 2001; Beer et al., 2002]. It is possible that it is because adenocarcinomas have been examined in the greatest numbers that subgroups of tumors with different clinical behaviors have been identified, as significant heterogeneity is not restricted to adenocarcinomas. These studies suggest however, a number of important points that could be the basis for more extensive analyses in the future. Firstly, by examining the gene expression of lung cancers it may be possible to not only define the genes that are unique and distinguish each of the various types or subtypes. Secondly, this information may also help determine the genes that may be in common, and help to determine the relationships between these tumors and their potential cells of origin. Thirdly, the analysis of a sufficiently large and robust set of tumors carefully annotated in regards to the clinical behavior of each tumor, as has been done for lung adenocarcinomas, may provide more insight into the biological processes that underlie these tumor metastatic and aggressive behavior.

Over the last four decades, the age-adjusted incidence of the different types of lung cancer [Zheng et al., 1994] indicates that among both men and women, a steady increase in all of the major types were observed except for bronchioloalveolar carcinomas, which remained relatively constant. In the mid 1980's squamous cell carcinomas and small cell cancers among men decreased slightly, whereas a steady increase was observed in these cancers among women. The incidence of adenocarcinomas has continued to increase in both sexes but this has increased more significantly among women. The basis for these trends is likely to be related to use of tobacco products, as cigarette smoking is the major risk factor for the development of lung cancer. Women have a 1.2 to 1.7 fold higher odds ratio of developing lung cancer of all types than men even though women tend to start smoking later and inhale less deeply than men [Zang et al., 1996]. This is a very important and interesting difference and may relate to a higher susceptibility due to nicotine metabolism, cytochrome P450 enzyme variations, or hormonal influences on tumor development. A study by Ryberg et al. [1994] suggested that DNA adduct levels are higher among women than men after adjusting for smoking dose. Therefore gene expression profiling may have the potential to uncover additional reasons for the gender-related differences in lung cancer susceptibility.

Although smoking is most strongly associated with squamous cell and small cell lung cancer, and historically adenocarcinomas have been most common among non-smokers, most patients who develop adenocarcinomas of the lung have a smoking history [Brownson et al., 1992]. Interestingly, the increased incidence in adenocarcinomas and the decrease in squamous and small cell lung cancers is thought to reflect the changed manufacturing from unfiltered to filtered cigarettes [Lubin et al., 1984; Wynder and Kabat, 1988]. At least one study by Miura et al. [2002] has compared the gene expression profiles of lung tumors among smokers and non-smokers. They found expression differences and abnormal expression of genes involved in spindle checkpoint and genomic stability in lung adenocarcinomas from smokers. These studies suggest that gene expression studies may not only be able to determine those differences that may be directly related to cigarette smoking but may also be able to provide biological insight into tumors associated with the other major risk factors for lung cancer. This will require accurate clinical histories and exposure data however. For example, occupational exposure to radon, asbestos, diesel engine exhaust, silica, arsenic, chromium, cadmium, nickel are all associated with an elevated risk of lung cancer [Rivera et al., 2001]. Careful analysis of the lung cancers associated with a given patient cohort may define previously unappreciated differences between tumors caused by different agents, or alternatively identify similar underlying mechanisms between various agents.

1.3 Gene Profiling as an Adjunct to Tumor Staging

The best predictor of patient outcome is tumor stage. Tumor staging is used to define groups of patients that are distinct from one another. The 1997 American Joint Commission for Cancer and the UICC defined the TMN classification system for tumor staging that is based on the definition of the anatomic extent of the disease [AJCC, 1998]. The T concept reflects both the size and location of the tumor, the N concept defines the lymph nodal involvement in terms of the tumor location, and the M concept defines the presence of distant metastasis. Stage I lung cancers may include Ia and Ib. Stage Ia tumors (T1N0M0) are small (< 3cm) with no nodal involvement and no evidence of distant metastasis. Stage Ib (T2N0M0) are larger (>3 cm) and may show pleural involvement or are located centrally in lobar bronchus. Stage II tumors include IIa and IIb. Stage IIa (T1N1M0) and IIb (T2N1M0) both involve nodes within the lung but differ in the tumor size. Stage IIIa (T1-3N2M0) can include tumors that differ in size and location in the lung but involve ipsilateral mediastinal lymph nodes. Stage IIIb (T1-3N3M0) involve the contralateral mediastinal or supraclavicular nodes. Stage IV tumors are those that are metastatic to distant sites (M1). Clinical stage refers to pretreatment and pathologic stage is following resection and detailed assessment of the lesion. The tools that may be utilized for staging include chest radiography (CXR), computed tomography (CT), magnetic resonance imaging (MRI), positron-emission tomography (PET), mediastinoscopy (sampling nodal stations), and thoracotomy with lymph node dissection.

The size of a tumor is used in lung cancer staging and an association of increasing tumor size with decreasing survival has been observed [AJCC, 1998]. The growth rate of lung tumors is a highly dynamic process that can vary during the natural history of each lesion and mathematical models have been developed to help understand tumor growth and plan proper treatment [Norton et al., 1976; Calderon et al., 1991]. The doubling times differs among the various cancer types with small cell lung cancers having relatively short doubling times of only 30 days, and non-small cell lung cancers having doubling times of approximately 100 days [Geddes et al., 1979]. Within the non-small cell lung tumors there are also differences as large cell and squamous tumors have a doubling time of 90 days whereas for adenocarcinomas it is approximately 160 days. There are large variations observed even within each subtype of lung cancer yet proliferation rates of lung cancers may be one measure of their potential sensitivity to chemotherapeutic agents but unfortunately not the only factor.

The nodal involvement of lung cancers also directly correlates with the survival of patients with this disease [AJCC, 1998]. Lung cancers that spread to the different nodes within the lung or mediastinum are defined either as N1, N2 or N3. N1 are nodes within the lung (stations 10-14), N2 nodes are ipsilateral mediastinal (stations 1-9) and N3 are nodes in the contralateral mediastinum (stations 2,4). The five-year survival of patients with no nodal involvement (N0) is approximately 60%, N1 is 40%, N2 below 20%, N3 is about 10% and with distant metastasis it is less than 5% [Naruke et al., 1993]. Lung tumors that do metastasize to distant sites are most often observed in decreasing order, in the liver, bone, lung, brain, adrenal and kidney [Notter and Schwegler, 1989].

Although tumor stage incorporates information regarding aspects of the size of the tumor and it's potential spread to lymph nodes or to distant sites, stage itself may be insufficient to fully explain the behavior of all lung cancer especially for early stage lung tumors. This may be due to the fact that definitive identification of every potential metastatic cell may not be possible. For example most tumors that are only 1-2 cm may not have metastasized, yet some may already have and even careful pathological examination of dissected lymph node might fail to identify a small number of disseminated cells. Thus as indicated in Table 1 there are many clinically important questions that might be addressable using gene expression profiling approaches including the following. Are lung tumors within the same pathological-based stage a homogenous group? Are lung tumors of the same histological type but differing in tumor size, or metastasize to N1, N2 or N3 positions different? Are tumors that show hematogeneous metastasis different from those that spread via the lymphatics? Are tumors that metastasize to different organ sites discernable? To accurately answer these questions it will require appropriately designed studies with sufficiently large numbers of well-annotated tumors with information regarding many of these critical biological properties.

1.4 Critical Issues Affecting Gene Expression Profiling

The challenge to the research community is to identify and interpret the important information that may be present in the gene profiles of lung tumors pertaining to the clinically relevant questions raised above. The extensive heterogeneity of lung cancer has been long appreciated especially from the analyses of genomic alterations detected using genomic and karyotype-based approaches [Luk et al., 2001]. These alterations undoubtedly underlie much of the observed heterogeneity at the mRNA or protein level. This heterogeneity is important since multiple areas of a tumor could differ in characteristics such as cell differentiation, drug responsiveness, tumor invasion or metastatic behavior. Thus an assumption is that the primary tumor used for mRNA isolation and gene expression analysis, will have at least some component that reflects the properties that are most clinically relevant. This may not always be the case, and if only a very small region of a given tumor is sampled for gene profiling analysis this potential confounding factor may become very important. In contrast to this, there is also the possibility that the primary tumor and the metastatic tumor cells are quite similar in regards to gene expression so that such a sampling issue is not a significant problem. Studies by Garber et al. [2001] have noted that when 918 genes are used for hierarchical clustering of lung tumors, both primary and metastatic tumors clustered immediately adjacent to one another in the clustering dendrogram. This would appear to suggest that although large number of genes detected by gene expression analyses that differ quite dramatically may be present, a subset of similarly expressed genes exist which is sufficient in identifying the relatedness of the metastatic cells and the original primary tumor.

The analysis of genes that correlate with patient survival may also be subject to variables that may complicate assessment of a direct involvement. For example, the age or the performance status of the patient with lung cancer may influence overall survival. A very elderly patient or one with a reduced performance status may not survive quite as long as a younger, healthier individual with a similarly aggressive cancer. Patients may also die for reasons unrelated to their disease but it may be difficult to separate these events. It is only by examination of a sufficiently large group of lung cancer patients that these potential complications may be minimized. Although survival associated genes representing important biological processes may be shared by lung tumors of diverse histological type or containing diverse genetic alterations, it is possible that there may be unique sets of genes that may be specific for individual or similar groups of lung cancers. For example, a somatic alteration resulting in the high level amplification of a gene such as a growth factor receptor may be a relatively rare event in lung cancer however, for individuals whose tumors demonstrate this alteration, this gene may be highly associated with the overall survival. This was observed for the *erb*B2 gene in the analysis of survival of lung adenocarcinomas [Beer et al., 2002], as were other outlier type genes that may reflect gene alterations that are more unique for individual patients tumors. Only by examination of a large number of tumors of similar type (i.e. adenocarcinomas) however, may it be possible to determine whether there are smaller subsets of recurring genes that are strongly influencing tumor behavior and patient survival.

1.5 Conclusions

The application of gene expression profiling to lung cancer has begun to provide large amounts of information that when properly interpreted may provide new insights into this disease that continues to represent a significant health problem. The types of questions that are being asked should influence the design of the studies. Because early detection may actually be one of the most effective mechanisms to increase the survival of patients with lung cancer, gene expression profiling studies that are focused on identifying highly expressed genes or those unique to specific lung subtypes are appropriate. The question becomes what are the correct comparison tissues for these analyses? Ideally the diagnostic genes will be highly expressed in the lung cancer and not expressed or only show low-level expression in the cells of origin in the normal lung. This is a difficult problem due to the many cell types present in the normal lung. Comparison of the normal bronchiolar squamous mucosa to squamous cell carcinomas would be ideal, yet few other comparisons are easily made for the other lung cancer types without using laser capture technologies.

Many studies to date have examined multiple types of lung tumors and identified unique sets of gene expression characteristics. Because of the potential variation in the analyses from one study to another and the use of separate gene analyses platforms, the pooling of these data is challenging, but may be useful for testing hypotheses generated from one individual set of lung tumors to another. The comparisons that may prove most useful for answering the many important, clinically relevant questions raised in this review will be when sufficient numbers of well-annotated tumors are compared to each other. To date lung adenocarcinomas have been the most extensively examined yet much larger sets of samples will be needed to more completely characterize this tumor type and address the most important questions. There is a need to provide a further understanding of the genes and processes that underlie a tumor's hematogenous or lymphatic route of metastasis or those genes that might predict this behavior even in early stage tumors. It is also hoped that these studies will also uncover new targets for therapeutic intervention or predict which patients may respond best to specific chemotherapeutic agents. In this latter aspect, the design of the studies requires annotation of the tumor's gene expression profile with the patient's treatment and response to specific agents. This will be most effectively accomplished in the setting of clinical trials and obtaining pretreatment biopsies will need to be incorporated in the protocols to allow such studies to be effective and informative. At present most studies that have examined gene expression of lung cancer are from patients with surgically resectable disease. Since most lung cancer patients present with advanced disease there will be the need to adapt analysis methods to use either pretreatment biopsies or cytological preparations from this patient population. This is especially important if we are to identify expression patterns that are associated with favorable or unfavorable response to specific therapeutic regiments. The potential of gene expression analyses to be utilized in the clinical setting may be on the horizon and could provide a basis for individualized patient therapy.

Gene expression analyses of lung cancer and the extraction of clinically relevant information using bioinformatics represent a promising yet challenging endeavor. Given the appropriate experimental design and attention to those factors that are most critical to providing gene expression data of high quality, these approaches have significant promise in helping to address many clinically important problems associated with lung cancer.

Table 1. Questions representing opportunities for investigation.

Are lung tumors within the same pathological-based stage a homogeneous group?
Are lung tumors of the same histological type but differing in tumor size, or metastasize to N1, N2 or N3 positions different?
Are the gene expression patterns of tumors that metastasize to different organ sites discernable?
What are the reasons for the gender-related differences in lung cancer susceptibility?
Are the survival-related genes among the different subtypes of lung cancer similar or different? What genes/profiles are associated with response or resistance to therapeutic agents?

1.6 References

American Joint Commission for Cancer. 1998, AJCC Cancer Staging Handbook. **5th** ed. Philadelphia, PA: Lippincott-Raven.

Beer, D.G., Kardia, S.L.R, Huang, C.C., Misek, D.E., Lin, L., Chen, G., Gharib, T.G., Giordano, T.J., Thomas, D.G., Lizyness, M.L., Kuick, R., Taylor, J.M.G., lannettonni , M.D., Orringer, M.B., Hanash, S., 2002, Gene expression profiles define a high-risk group among stage I lung adenocarcinomas. Nat Med. **8**:816-824.

Bhattacharjee, A., Richards, W.G., Staunton, J., Li, C., Monti, S., Vasa, P., Ladd, C., Behesti, J., Bueno, R., Gillette, M., Loda, M., Weber, G., Mark, E.J., Lander, E.S., Wong, W., Johnson, B.E., Golub, T.R., Sugarbaker, D.J., Meyerson, M., 2001, Classification of human lung carcinomas by mRNA expression profiling reveals distinct adenocarcinoma subclasses. Proc Natl Acad Sci. (USA) **98**:13790-13795.

Borczuk, A.C., Gorenstein, L., Walter, K.L., Assaad, A.A., Wang, L., Powell, C.A., 2003 Non-small cell lung cancer molecular signatures recapitulate lung developmental pathways. Am J Pathol. **163**:1949-1960.

Brambilla, E., Travis, W.D., Colby, R.V., Corrin, B., Shimosato, Y., 2001, The new World Health Organization classifiction of lung tumours. Eur Respir J. 18(6): 1059-1068.

Brownson, R.C., Chang, J.C., Davis, J.R., 1992, Gender and histologic type variations in smoking-related risk of lung cancer. Epidemiol. **3**:61-64.

Bulzebruck, H., Bopp, R., Drings, P., Bauer, E., Krysa, S., Probst, G., van Kaick, G., Muller, K.M., Vogt-Moykopf, I., 1992, New aspects in the staging of lung cancer: prospective validation of the International Union Against Cancer TNM classification. Cancer **70**:1102-1110.

Calderon, C., Kwembe, T., 1991, Modeling tumor growth. Math Biosci. **103**(1):97-114.

Difilippantonio, S., Chen, Y., Pieta, A., Schluns, K., Pacyna-Gengelbach, M., Deutschmann, N., Padillla-Nash, H.M., Ried, T., Peterson, I., 2003, Gene expression profiles in human non-small and small cell lung cancers. Eur J Oncol. **39**: 1936-1947.

Ferlay, J., Bray, F., Pisani, P., Parkin, D.M., 2001, Globocan 2000: Cancer Incidence, Mortality and Prevalence. Lyon, IARC Press.

Garber, M.E., Troyanskaya, O.G., Schleens, K., Petersen, S., Thaesler, Z., Pacyna-Gengelbach, M., van de Rijn, M., Rosen, G.D., Perou, C.M., Whyte, R.I., Altman, R.B.,

Brown, P.O., Botstein D, Petersen, I., 2001, Diversity of gene expression in adenocarcinoma of the lung. Proc Natl Acad Sci. (USA) **98**:13784-13789.

Geddes, D.M., 1979, The natural history of lung cancer: a review based on rates of tumor growth. Br J Dis Chest **73**:1-17.

Heighway, J., Knapp, T., Boyce, L,, Brennand, S., Field, J.K., Betticher, D.C., Ratschiller, D., Gugger, M., Donovan, M., Lasek, A., Pickeert, P., 2002, Expression profiling of primary non-small lung cancer for target identification. Oncogene **21**:7749-7763.

Jemal, A., Murray, T., Samuels, A., Ghaffor, A., Ward, E., Thun, M.J., 2003, Cancer statistics, 2003. CA Cancer J Clin. **53**:5-26.

Kikuchi, T., Daigo, Y., Katagiri, T., Tsunoda, T., Okada, K., Kakiuchi, S., Zembutsu, H., Furukawa, Y., Kawamura, M., Koichi, K., Imai, K., Nakamura, Y., 2003, Expression profiles of non-small cell lung cancers on cDNA microarrays: Identification of genes for prediction of lymph-node metastasis and sensitivity to anti-cancer drugs. Oncogene **22**:2192-2205.

Lubin, J.H., Blot, W.J., Berrino, F., Flamant, R., Gillis, C.R., Kunze, M., Schmahl, D., Visco, G., 1984, Patterns of lung cancer risk according to type of cigarette smoked. Int J Cancer **33**:569-576.

Luk, C., Tsao, M.S., Bayania, J., Sheppard, F., Squire, J.A., 2001, Molecular cytogenetic analysis of non-small cell lung carcinoma by spectral karyotyping and comparative genomic hybridization Cancer Genet Cytogenet. **125**:87-99.

McDoniels-Silvers, A.L., Nimri, C.F., Stoner, G.D., Lubet, R.A., You, M., 2002. Differential gene expression in human lung adenocarcinomas and squamous cell carcinomas. Clin Cancer Res. **8**:1127-1138.

Miuri, K., Bowman, E.D., Simon, R., Peng, A.C., Robles, A.I., Jones, R.T., Katagiri, T., He, P., Mizukami, H., Charboneau, L., Kikucki, T., Liotta, L.A., Nakamura, Y., Harris, C.C. , 2002, Laser capture microdissection and microarray expression analysis of lung adenocarcinoma reveals tobacco smoking- and prognosis-related molecular profiles. Cancer Res. **62**:3244-3250.

Nacht, M., Dracheva, T., Gao, Y., Fujii,T., Chen, Y., Player, A., Akmaev, V., Cook, B., Dufault, M., Zhang, M., Zhang, W., Guo, M-Z., Curran, J., Han, S., Sidransky, D., Buetow, K., Madden, S.L., Jen, J., 2001, Molecular characteristics of non-small cell lung cancer. Proc Natl Acad Sci. (USA) **98**:15203-15208.

Nakamura, H., Saji, H., Ogata, A., Hosaka, M., Hagiwara, M., Saijo, T., Kawasaki, N., Kato, H., 2003, cDNA microarray analysis of gene expression in pathologic stage IA non-small cell lung carcinomas. Cancer **97**:2798-2805.

Naruke, T., 1993, Significance of lymph node metastasis in lung cancer. Semin Thorac Cardiovasc Surg. **5**:210-218.

Norton, L., Simon, R., Bereton, H., Bodgen, A., 1976, Predicting the course of Gompertzian growth. Nature **264**:542-545.

Notter, M., Schwegler, N., 1989, Incidence of metastases and their distribution pattern in bronchial carcinoma. Autopsy findings during 5 decades (1935 to 1984) Dtsch Med Wochenschr. **114**(9):343-349.

Petersen, S., Heckert, C., Rudolf, J., Schlun, K., Tchernitsa, O.I., Schafer, R., Dietel, M., Petersen, I., 2000, Gene expression profiling of advanced lung cancer. Int J Cancer **86**:512-517.

Rivera, M.P., Detterbeck, F.C., Loomis, D.P., 2001, Epidemiology and classification of lung cancer, In: Diagnosis and Treatment of Lung Cancer: An Evidence-based Guide for the Practicing Clinician, W.B. Saunder Co, Philadelphia PA, Chapter 3, pp: 25-44.

Ryberg, D., Hewer, A., Phillips, D.H., Haugen, A., 1994, Differential susceptibility to smoking-induced DNA damage among male and female lung cancer patients. Cancer Res. **54**:5801-5803.

Travis, W.D., Travis, L.B., Devesa, S.S., 1995, Lung cancer. Cancer **75**:191-202.

US Dept. of Health and Human Services, 1998, Cigarette smoking among high school students-11 states, 1991-1997. Morbid Mortal Weekly Rep. **48**:686-692.

Wigle, D.A., Jurisica, I., Radulovich, N., Pintilie, M., Rossant, J., Liu, N., Lu, C., Woodgett, J., Seiden, I., Johnston, M., Keshavjee, S., Darling, G., Winton, T., Breitkreutz, B-J., Jorgenson, P., Tyers, M., Sheppard, F.A., Tsao, M.S., 2002, Molecular profiling of non-small cell lung cancer and correlation with disease-free survival. Cancer Res. **62**:3005-3008.

Wikman, H., Kettunen, E., Seppanen, J.K., Karajainen, A., Hollmen, J., Anttila, S., Knuutila, S., 2002, Identification of differentially expressed genes in pulmonary adenocarcinoma by using cDNA array. Oncogene **21**:5804-5813.

Wynder, E.L., Kabat, G.C., 1988, The effect of low tar cigarette smoking on lung cancer risk. Cancer **62**:1223-1230.

Yamagata, N., Shyr, Y., Yanagisawa, K., Edgerton, M., Dang, T.P., Gonzalez, A., Nadaf, S., Larsen, P., Roberts, J.R., Nesbitt, J.C., Jensen, R., Levy, S., Moore, J.H., Minna, J.D., Carbone, D.P., 2003, A training-testing approach to the molecular classification of resected non-small cell lung cancer. Clin Cancer Res. **9**:4695-4704.

Zheng, T., Holford, T.R., Boyle, P., Chen, Y., Ward, B.A., Flannery, J., Mayne, S.T., 1994, Time trend and the age-period-cohort effect on the incidence of histologic types of lung cancer in Connecticut, 1960-1989. Cancer **74**:1556-1567.

Zang, E.A., Wynder, E.L., 1996, Differences in lung cancer risk between men and women: examination of the evidence. J Natl Cancer Inst. **88**:183-192.

Chapter 2

GENE EXPRESSION DATA AND SURVIVAL ANALYSIS

Peter J. Park
Children's Hospital Informatics Program and Harvard-Partners Center for Genetics and Genomics, NRB 255, 77 Avenue Louis Pasteur, Boston, MA 02115

Abstract: Finding associations between expression profiles and simple phenotypic data such as class labels has been studied extensively, including prediction algorithms for new samples based on these relationships. However, much work is needed to link expression profiles to more complex response variables, most notably survival data with censoring. Reducing the survival data to a short-term versus long-term survival indicator or using survival curves merely to demonstrate the difference between clusters of samples is not an efficient use of the data. We review some of the progress and challenges in this area. We discuss the need for more consistent results among studies done on different microarray platforms, for development of sample-specific predictive scoring schemes, and for a more comprehensive analysis that incorporates other prognostic factors and clearly demonstrates the added value of expression profiling over current protocols.

Key words: Cluster analysis; dimensionality reduction; censored data; Kaplan-Meier analysis; cross-platform comparisons

1. INTRODUCTION

From the beginning, one of the most exciting areas of application envisioned with the microarray technology has been its use in the clinic. By obtaining a 'molecular portrait' of diseases, we would gain fresh understanding of the disease processes at the molecular level, which would allow us to improve our classification of diseases and aid in discoveries of new subtypes. This would quickly lead, it was advertised by some, to the realization of 'personalized medicine,' in which diagnosis and prognosis, as well as treatment plan, would depend on the individual's genetic information.

In the past few years, there has been a great effort in many aspects of this endeavor, to a varying degree of success. Although much is left to be desired, there has been substantial improvement in the quality of the transcript measurements, both for spotted cDNA arrays and for oligonucleotide arrays by Affymetrix. There have been some alternative technologies as well, especially the spotted or printed oligonucleotide arrays with longer probes. In terms of analysis, much of the initial work has been in associating expression data with a binary response variable such as the labels indicating cancer or normal tissue. The common tasks have been to identify genes that are highly correlated with the disease classification and then to use these genes to build a prediction scheme. Numerous methods of varying complexity have been applied to this problem. Starting with the 'signal-to-noise' metric and 'weighted-voting' prediction scheme in Golub et al. [1999], a seemingly countless number of methods from numerous disciplines has been applied to this problem, ranging from traditional statistical techniques to the latest computer-intensive techniques. Unfortunately, it is still unclear which method performs the best in general because too many methods have been applied to few relatively easy datasets, all claiming superiority against a method known to be less than optimal. A subset of these methods was subsequently expanded to the case of multiple classes, in order to deal with many subtypes of diseases [Bhattacharjee et al., 2001; Ramaswamy et al., 2001; Pomeroy et al., 2002; Rifkin et al., 2003]. Many modifications to the multi-class problem, however, have been relatively simple extensions of the binary case, in which a series of one versus many comparisons are combined.

There are other types of data besides these nominal ones that will be important in more comprehensive studies in the future. In ordinal data, the order is important in the categories, such as 'minor', 'moderate', 'severe', or 'fatal' for a disease progression; in discrete data, both the order and the magnitude are important, such as the number of relapses of a disease; and in continuous data, the measurements are not restricted to specified values. Effective methods are yet to be determined or developed in most of these cases.

An important phenotypic variable which has received more attention recently is the patient survival times. It has been gradually recognized that gene expression must be considered in the context of all other patient characteristics and that it can provide more information than simple disease classification. Survival times are obviously an important characteristic that has direct and immediate implications. Survival analysis is a collection of statistical methods for describing the distribution of survival/failure times or any other time-to-event, and a large number of tools exist in that literature. However, there has not been a strong link yet to the high-dimensional setting encountered with expression data. Some properties and methods of analysis for survival data will be described briefly in the next section, but it suffices to note that the number of studies correlating gene expression with survival data has increased dramatically. Well represented already are different types of cancers, but now there are studies involving other clinical aspects, such as renal allograft rejection [Sarwal et al., 2003].

With survival data, careful analysis is imperative [Altman and Royston, 2000]. For example, many earlier studies may be flawed in their claims of high prediction rate in classification of samples due to a selection bias [Ambroise and McLachlan, 2002; Simon et al., 2003]. A simple simulation in Simon et al. [2003] shows that a high classification rate can be achieved in the popular leave-one-out cross-validation even for randomly generated data if the sample that is left out for testing has been used in generating the list of genes used in prediction. While this may appear obvious, the mistake happens surprisingly often. It is common, for example, to normalize the data and filter the genes to get a manageable number of genes, in the order of few hundred or a thousand genes, before the main analysis including cross-validation is carried out. The estimation of correct leave-one-out prediction rate, however, requires that the whole process including normalization and filtering be repeated each time a sample is left out. The effect of normalization while including the validation sample may not be large, but the effect of a filtering in the same way is often larger than expected. This error is no longer prevalent in the literature, but other subtle issues still remain. In some instances, the estimation of the prediction rate involves some circularity, with genes used as predictors having been used in the first place to define the groups [Sorlie et al., 2003].

2. CURRENT USE OF SURVIVAL DATA

The main difficulty with patient survival data is the presence of censoring. Censoring occurs when the outcome is not observed for a patient. For example, in a cancer trial, a group of patients is followed prospectively for a period of time and the outcome variable may be time to death. At the end of the study period, however, mostly likely not all patients will have died; also, some patients may have left the study early for reasons unrelated to the disease or trial. We may know that a patient has lived at least two years, for instance, but do not know the exact time of death. Death is one example of an event; in general, the response variable can be any time-to-event data. Common variables include time to relapse of a disease, number of occurrences of a disease, and time to disease-free state after a treatment. Censoring can be either left-censored or right-censored or both. In a prospective study, a group of patients with a particular disease may be recruited and followed. For some patients, the date of diagnosis may be known but they may be considered left-censored if the disease was contracted at some unknown time prior to the diagnosis. Unless this effect is severe or can be corrected in the analysis, however, we assume that the time of diagnosis is close to the start of disease and do not consider them as censored. Right-censoring is generally the more serious problem and cannot be ignored if unbiased estimates are to be obtained. We usually assume uninformative censoring, that the censoring is not related to the effects under investigations. For example, if a patient drops out of a study because he has moved to another location, that is considered uninformative; if he drops out due a deteriorating condition, that is not uninformative. Without this assumption, the analysis becomes more difficult if not impossible.

In most clinical studies, censoring is a serious issue that must be dealt with efficiently. Observed endpoints are desirable from the analysis point of view, but censoring inevitably occurs; it is not unusual to have more than half the patients censored in a trial. In gene expression studies, survival endpoints have not been used in an efficient manner so far. In the simplest approach, patients were roughly divided into two categories, for short-term and long-term survival. This reduced the problem into the dichotomous variable case, for which numerous methods are available. The problem with this, however, is that much information is thrown away, as may be evidenced in large within-group heterogeneity. If a two-year survival is used as a cut-off in a study, for example, a patient who survived just over two years may be put in the same group as someone who survived ten years while he is put in the different group from someone who survived just under two years.

In a more popular approach, many of the studies employed the strategy of clustering the patients according to their expression profiles first and then showing that the patients in different clusters have statistically significant differences in survival outcomes. Hierarchical clustering has been the favored clustering scheme and using Kaplan-Meier curves and log-rank tests for patient survival have been the common methods for demonstrating the differences among clusters so far. The Kaplan-Meier method is a nonparametric technique for estimating the probability that an individual survives beyond a given time; the idea behind the log-rank test is to construct a series of contingency tables for group versus survival status at each time at which a failure occurs and then to combine the information from the tables using the Mantel-Haenszel statistic.

While this approach has been fruitful in demonstrating that there is a relationship between expression profiles and survival, it is an *indirect* and inefficient use of the data. Prediction is made only in that a patient may be put in one group which, on the average, has a different survival function from the others. In a sense, survival data are used merely to verify the effectiveness of the clustering algorithm. In fact, further separation in the Kaplan-Meier curve has been used as a criterion for judging the quality of clustering algorithms at times. A more effective use of the data would be to build a predictive model that will make direct profile-specific estimates on a continuous scale. There have been some progress in this direction and some examples are mentioned in the next section.

3. CHALLENGES

3.1 Technological limitations

One of the major problems in expression analysis has been the lack of consistency and reproducibility in the data. If, for instance, the relationship between an expression profile and its survival prediction holds only within a particular study, it is not clear how much of the conclusion was real and how much was an artifact of the analysis method. Before microarrays can be used more routinely for diagnostic or prognostic purposes, this issue of reproducibility must be better understood.

Some have suspected early on that there may be substantial difference in the results from the cDNA arrays manufactured in small-scale laboratories and from the oligonucleotide arrays, especially the high-density Affymetrix arrays with multiple 25-mer probes for each target sequence. This lack of concordance was first reported in Kuo et al. [2002] using the data from a panel of 60 cancer cell lines from the National Cancer Institute that were hybridized onto both cDNA and Affymetrix arrays. Subsequently, there have been other studies describing similar discordance for platforms including the spotted or printed oligo arrays [Yuen et al., 2002; Tan et al., 2003].

This issue is serious even within the same platform. Much of the work with clinical application has been done with Affymetrix arrays, starting with HuFL arrays and continuing with U95A-E, followed by U133A-B and U133 2.0 Plus series. However, while the basic fabrication technology has stayed the same, there have been important differences among the different generations of arrays, for example, with different number of probes in a probe set for each gene. Improvements have come most notably in the probe selection algorithms and in the calculation of summary expression measures.

In Nimgaonkar et al. [2003], it is observed that there is substantial disagreement among the Unigene matched genes between the HuFL and U95A platforms, with greater agreement when more probes are shared between the two probe sets for a gene. We have recently carried out a more extensive and quantitative comparison using a dataset in which each sample was hybridized both to U95A and to U133A (manuscript in submission). With several commonly used methods of matching genes across the arrays and preprocessing them, we were unable to reduce the dominant effect of the array type: an unsupervised clustering results in a separation by array type rather than disease type and there is substantial difference in the genes identified as differentially expressed.

Given the difficulties of comparing data even among succeeding generations of arrays within the same technology platform, it is not surprising that data generated with differences in samples, instruments, institutions, protocols, and platforms do not agree. Three prominent cases of diseases with multiple data sets are diffuse large B-cell lymphoma [Alizadeh et al., 2000; Rosenwald et al., 2002; Shipp et al., 2002]; lung carcinoma [Bhattacharjee et al., 2001; Garber et al., 2001; Beer et al., 2002; Wigle et al., 2002]; and breast cancer [Perou et al., 2000; Hedenfalk et al., 2001; Sorlie et al., 2001; West et al., 2001; van de Vijver et al., 2002; van 't Veer et al., 2002; Huang et al., 2003]. While the general conclusions of these studies are the same, specific results can vary substantially. In particular, the overlap of the marker gene lists is surprisingly small in general [Sorlie et al., 2003].

Another reason for the disagreement in the results is the different algorithms and their lack of robustness in data analysis. In Sorlie et al. [2001], genes useful for classification were determined using patient survival as the supervising variable in Significance Analysis of Microarrays [Tusher et al., 2001]; in Jenssen et al. [2002], the same dataset was analyzed using a variation on the univariate log-rank test on each gene. However, only 29 genes were common between the two lists containing 264 and 95 genes. Compared to a different data set [van 't Veer et al., 2002] that was analyzed with the occurrence of metastasis as the patient outcome, only two genes were in common between the lists with 174 and 95 genes.

This is in some respects reminiscent of the lack of reproducibility in association studies that look for common genetic variants, such as single nucleotide polymorphisms (SNPs), that contribute to disease susceptibility. In these studies, most associations claimed do not appear to be robust [Hirschhorn et al., 2002; Lohmueller et al., 2003]. In one study, a meta-analysis showed that of the 166 putative associations which have been studied three or more times, only six have been consistently replicated [Hirschhorn et al., 2002]. The underlying problem is similar in both expression studies and genetic association studies: there are many factors that contribute the phenotype but each with only a modest contribution. Well-controlled studies with large samples and robust analysis are needed to verify the results in both cases.

There has been at least some success in comparing independent data sets. In Sorlie et al. [2003], class prediction using a variant of nearest-centroid classification method [Tibshirani et al., 2002] was performed, with one dataset as a training set and two independent datasets as testing sets. Although the number of genes shared in the informative gene lists was small and using a set of marker genes found in one study only to predict the outcomes in another does not perform well, using a set of common markers performed better and similar subtypes were observed in all cases [Sorlie et al., 2003].

3.2 Dealing with high-dimensionality

Any correlative analysis of survival data with gene expression inherits all the problems associated with high-dimensional datasets in addition to the problems caused by censoring. This is a fundamentally difficult problem, and there are no simple solutions that are both mathematically rigorous and offer biologically meaningful interpretation. A large part of current analysis consists of exploratory analysis based on experience and available software.

After some initial filtering to eliminate non-expressed genes and genes with small variability, the next step is to further reduce the number of predictors to find informative genes. One approach is to use a well-known mathematical technique for dimensionality reduction. Principal component analysis and singular value decomposition are typical methods in this category [Alter et al., 2000]. While mathematically attractive, the two main disadvantages of these are that principal components or singular vectors may not highly correlated with the outcome variable and that it is difficult to assign meaning to them except in few simple cases. Sometimes the coefficients of principal components or singular vectors can be examined to determine those genes with large contributions, but usually a small subset of dominant genes does not exist. If the goal of an expression profiling project is to simply devise the most accurate prediction scheme regardless of its interpretability, such a dimensionality reduction method followed by a machine learning technique may give good results [Khan et al., 2001]. There are many machine learning methods such as neural networks and genetic algorithms, but support vector machines appear to perform especially well for that purpose [Brown et al., 2000].

For a more complex analysis involving survival phenotype, a better approach is to reduce the number of variables by grouping genes that are similar in some measure, creating what some have referred to 'metagenes' or 'supergenes.' There are many variations on this idea. Based on Tukey's idea of compound covariates [Tukey, 1993], one can form a linear combination of genes with similar expressions with weights corresponding to the statistic from the two-sample t-test. This is the approach taken in Hedenfalk et al. [2001] and discussed further in Radmacher et al. [2002], including its relationship to the weighted-voting method [Golub et al., 1999]. In the tree-harvesting method [Hastie et al., 2001], a step-wise regression is used to select gene clusters of varying sizes that are related to the phenotype, based on the Cox proportional hazard model. The clusters may be derived, for example, from hierarchical clustering. In Rosenwald et al. [2002], Cox proportional hazard model was apply on individual genes and these were clustered into 'signature groups.' From this a smaller set was chosen as representative genes and averaged values of similar genes were included in a multivariate Cox model. The model was then used to compute a risk score for each patient. In this work, gene annotations were considered in grouping of the genes in addition to expression similarities and, as a result, the new reduced set of variables provides convenient biological interpretations. Multivariate Cox model was also fit in van de Vijver et al. [2002], but in this work expression profiles was reduced to an indicator variable as 'good-prognosis' versus 'bad-prognosis' signature. There are other methods of grouping genes for prediction, such as model-based clustering of genes [McLachlan et al., 2002]. In all these methods, the goal is to reduce the number of genes in a reasonable manner such that a conventional tool for survival analysis such as the multivariate Cox model can be used. A model with biological interpretation such as in Rosenwald et al. [2002] appears especially helpful.

For the purpose of prediction, approaches based on partial least squares have been explored with considerable promise. While principal component analysis has been popular in an unsupervised setting, principal components capture the variability in the gene expression space only and may not be highly correlated with the response variable. On the other hand, variable selection in linear regression chooses genes that are highly correlated with the response variable but do not account for the variability in the gene space. Partial least squares lies in between, producing a set of orthogonal linear combinations of genes that are predictive of the response while capturing the variability in the predictor space. The popularity of partial least squares has been due to its adaptability in the presence of a large number of variables, even when it exceeds the number of cases. It appears to work well when the number of predictors exceeds the number of cases moderately, but it is not clear how scalable this result is for extreme cases. This approach has been applied in gene expression analysis [Nguyen and Rocke, 2002a; Johansson et al., 2003; Perez-Enciso and Tenenhaus, 2003] for nominal phenotype. For the censored phenotype, partial least squares was used as a dimension reduction tool [Nguyen and Rocke, 2002b]. In Park et al. [2002], the partial least squares approach was reformulated to an equivalent problem in the generalized linear regression setting for which partial least squares was already worked out in Marx [1996]. This formulation circumvents the issue of censored data, at the cost of increased dimension in the problem, and appears to perform very well. A related approach is based on kernel Cox regression models [Li and Luan, 2003] in the framework of support vector machines. This method is based on the reformulation of support vector machine as a penalization method in function estimation, with the negative partial likelihood in the Cox model as the loss function. As in partial least squares, a large number of genes may be included in the set of potential predictors.

3.3 Incorporating other patient data

While there has been much progress in showing association between expression profiles and disease subtypes or even patient survival, its practical value in the clinical setting over current protocols and guidelines has not been demonstrated as convincingly in most instances. This may be one reason for the absence of microarray experiments in the clinic at this point, even after numerous studies over many years claiming the usefulness of expression profiling.

There are several reasons for this. The first is that much initial work may not have been as practical as they might have appeared at first. In some cases, it is not surprising that certain types of tumors can be distinguished with expression data, since gene expression simply reflects the features of the different cell type or other underlying characteristics. Sometimes the main distinguishing feature of expression profiles in different groups may reflect a mutation in a gene, such as in breast cancer, in which case a screening for the mutation directly would be more cost-effective and as accurate. A main conclusion of Hedenfalk et al. (2001), for example, was that mutations of BRCA1 and BRCA2 influence the expression of a group of genes. In other cases, expression profiling has not been shown conclusively to be better than immunohistochemical staining that are easier to perform and less expensive.

Even when gene expression patterns are useful for classification of disease types and stages, only a small number of such studies have demonstrated that it is superior to current set of clinical parameters. Often, to show clinical relevance, a subclass of patients considered to be in the same stage of a disease under a standard protocol is shown to have varying survival times depending on their expression profiles. Substantial heterogeneity within a same category implies that further refinement using expression may be desirable. For example, in the case of diffuse large B-cell lymphoma patients, those with similar International Prognostic Index (IPI) still showed significantly different Kaplan-Meier curves when classified based on their expression profiles [Rosenwald et al., 2002]. In breast cancer, those patients with the same lymph-node status or risk group status showed significant differences in expression profiles with respect to both metastasis-free period and overall survival [van de Vijver et al., 2002]. This evidence of heterogeneity within a same group provides strong evidence; however, it may be, for example, that heterogeneity also exists in terms of the existing clinical parameters within those grouped by expression profiles.

To clearly demonstrate that expression profiling indeed contributes to a better classification and prediction after the effect of other prognostic factors are accounted, a multivariate model with other potential predictors should be considered. This was done, for example, in van de Vijver et al. [2002], although expression signature is entered only as an indicator variable in that work. Another method is to have a large enough cohort within a single stratum of patients with similar characteristics. In metastatic renal cell cancer, the current prognostic indicators are stage, grade, and Eastern Cooperative Oncology Group status and among stage IV tumors, no clinical parameters exist for predicting time to failure [Vasselli et al., 2003]. By considering only similarly staged patients with no other known prognostic indicator, clinical relevance of expression profiling was evidenced clearly in Vasselli et al. [2003]. More careful, integrative analysis in similar directions would be an important contribution.

4. CONCLUSIONS

We have briefly reviewed the use of survival data in the context of analyzing and utilizing gene expression data. While distinct expression profiles have been correlated with many disease types and this has resulted in much insight into the biological mechanism underlying these diseases, a more direct way to demonstrate their usefulness for patient care is by linking them to patient survival. Much of the work so far, however, has not utilized the survival data efficiently. In some cases, the survival data were used simply to divide the patient samples into short-term and long-term survival groups, so that previous methodologies for binary classification and prediction can be used; in other cases, the survival information was used merely to demonstrate that the groups obtained through a clustering method were sufficiently different. In order to take full advantage of the expression data, it is important to develop new statistical methodologies that are suited for survival data analysis in the high-dimensional setting, especially in the area of effective prediction algorithms involving censored data. Simple validation techniques also need to be developed, similar to the *n*-fold cross-validation approach that dominates the expression literature. It is also important to carry out careful analysis to demonstrate not simply that expression profiles are correlated with survival but that they are valuable when added to the information contained in more mundane covariates and currently available prognostic factors.

We have focused mostly on dealing with the case of censored response variable here, but there are more difficult cases that will become important in future studies. In some cases, the questions have been addressed already in the survival analysis or the clinical trials literature and need to be modified for use with genomic data; in other cases, new methods need to developed to answer fresh questions. When microarrays become part of longitudinal studies, methodologies will be needed to deal with repeated measurements [Laird and Ware, 1982]. Also, there may be more than a single phenotypic response in future studies and methods for multivariate response variables need to be studied. Incorporating other genomic data other than microarrays effectively also remains an issue. Finally, it is important in these cases that the more traditional statistical approach with emphasis on model building followed by model-checking with residuals and outlier detection needs to be reconciled with the more algorithmic approach driven by prediction rates and functional minimization from the computer science community.

5. ACKNOWLEDGEMENTS

I would like to thank the three anonymous referees for their careful reading of the manuscript and helpful suggestions.

6. REFERENCES

Alizadeh AA, Eisen MB, Davis RE, Ma C, Losses IS, Rosenwald A, Boldrick JC, Sabet H, Tran T, Yu X, Powell JI, Yang L, Martl GE, Moore T, Hudson J, Lu L, Lewis DB, Tibshirani R,Sherlock G, Chan WC, Greiner TC, Weisenburger DD, Armitage JO, Warnke R, Levy R, Wilson W, Grever MR, Byrd JC, Botstein D, Brown PO, Staudt LM.

(2000) Distinct types of diffuse large B-cell lymphoma identified by gene expression profiling. Nature 403:503-11

Alter O, Brown PO, Botstein D (2000) Singular value decomposition for genome-wide expression data processing and modeling. Proc Natl Acad Sci U S A 97:10101-6

Altman DG, Royston P (2000) What do we mean by validating a prognostic model? Stat Med 19:453-73

Ambroise C, McLachlan GJ (2002) Selection bias in gene extraction on the basis of microarray gene-expression data. Proc Natl Acad Sci U S A 99:6562-6

Beer DG, Kardia SL, Huang CC, Giordano TJ, Levin AM, Misek DE, Lin L, Chen G, Gharib TG, Thomas DG, Lizyness ML, Kuick R, Hayasaka S, Taylor JM, Iannettoni MD, Orringer MB, Hanash S (2002) Gene-expression profiles predict survival of patients with lung adenocarcinoma. Nat Med 8:816-24

Bhattacharjee A, Richards WG, Staunton J, Li C, Monti S, Vasa P, Ladd C, Beheshti J, Bueno R, Gillette M, Loda M, Weber G, Mark EJ, Lander ES, Wong W, Johnson BE, Golub TR, Sugarbaker DJ, Meyerson M (2001) Classification of human lung carcinomas by mRNA expression profiling reveals distinct adenocarcinoma subclasses. Proc Natl Acad Sci U S A 98:13790-5

Brown MP, Grundy WN, Lin D, Cristianini N, Sugnet CW, Furey TS, Ares M, Jr., Haussler D (2000) Knowledge-based analysis of microarray gene expression data by using support vector machines. Proc Natl Acad Sci U S A 97:262-7

Garber ME, Troyanskaya OG, Schluens K, Petersen S, Thaesler Z, Pacyna-Gengelbach M, van de Rijn M, Rosen GD, Perou CM, Whyte RI, Altman RB, Brown PO, Botstein D, Petersen I (2001) Diversity of gene expression in adenocarcinoma of the lung. Proc Natl Acad Sci US A 98:13784-9

Golub TR, Slonim DK, Tamayo P, Huard C, Gaasenbeek M, Mesirov JP, Coller H, Loh ML, Downing JR, Caligiuri MA, Bloomfield CD, Lander ES (1999) Molecular classification of cancer: class discovery and class prediction by gene expression monitoring. Science 286:531-7

Hastie T, Tibshirani R, Botstein D, Brown P (2001) Supervised harvesting of expression trees. Genome Biol 2:RESEARCH0003

Hedenfalk I, Duggan D, Chen Y, Radmacher M, Bittner M, Simon R, Meltzer P, Gusterson B, Esteller M, Kallioniemi OP, Wilfond B, Borg A, Trent J (2001) Gene-expression profiles in hereditary breast cancer. N Engl J Med 344:539-48

Hirschhorn JN, Lohmueller K, Byrne E, Hirschhorn K (2002) A comprehensive review of genetic association studies. Genet Med 4:45-61

Huang E, Cheng SH, Dressman H, Pittman J, Tsou MH, Horng CF, Bild A, Iversen ES, Liao M, Chen CM, West M, Nevins JR, Huang AT (2003) Gene expression predictors of breast cancer outcomes. Lancet 361:1590-6

Jenssen TK, Kuo WP, Stokke T, Hovig E (2002) Associations between gene expressions in breast cancer and patient survival. Hum Genet 111:411-20

Johansson D, Lindgren P, Berglund A (2003) A multivariate approach applied to microarray data for identification of genes with cell cycle-coupled transcription. Bioinformatics 19:467-73

Khan J, Wei JS, Ringner M, Saal LH, Ladanyi M, Westermann F, Berthold F, Schwab M, Antonescu CR, Peterson C, Meltzer PS (2001) Classification and diagnostic prediction of cancers using gene expression profiling and artificial neural networks. Nat Med 7:673-9

Kuo WP, Jenssen TK, Butte AJ, Ohno-Machado L, Kohane IS (2002) Analysis of matched mRNA measurements from two different microarray technologies. Bioinformatics 18:405-12

Laird NM, Ware JH (1982) Random-effects models for longitudinal data. Biometrics 38:963-74

Li H, Luan Y (2003) Kernel Cox regression models for linking gene expression profiles to censored survival data. Pac Symp Biocomput:65-76

Lohmueller KE, Pearce CL, Pike M, Lander ES, Hirschhorn JN (2003) Meta-analysis of genetic association studies supports a contribution of common variants to susceptibility to common disease. Nat Genet 33:177-82

Marx B (1996) Iteratively reweighted partial least squares estimation for generalized linear regression. Technometrics 38:374-381

McLachlan GJ, Bean RW, Peel D (2002) A mixture model-based approach to the clustering of microanay expression data. Bioinformatics 18:413-22

Nguyen DV, Rocke DM (2002a) Multi-class cancer classification via partial least squares with gene expression profiles. Bioinformatics 18:1216-26

Nguyen DV, Rocke DM (2002b) Partial least squares proportional hazard regression for application to DNA microarray survival data. Bioinformatics 18:1625-32

Nimgaonkar A, Sanoudou D, Butte AJ, Haslett JN, Kunkel LM, Beggs AH, Kohane IS (2003) Reproducibility of gene expression across generations of Affymetrix microarrays. BMC Bioinformatics 4:27

Park PJ, Tian L, Kohane IS (2002) Linking gene expression data with patient survival times using partial least squares. Bioinformatics 18 Suppl 1:S120-7

Perez-Enciso M, Tenenhaus M (2003) Prediction of clinical outcome with microarray data: a partial least squares discriminant analysis (PLS-DA) approach. Hum Genet 112:581-92

Perou CM, Sorlie T, Eisen MB, van de Rijn M, Jeffrey SS, Rees CA, Pollack JR, Ross DT, Johnsen H, Akslen LA, Fluge O, Pergamenschikov A, Williams C, Zhu SX, Lonning PE, Borresen-Dale AL, Brown PO, Botstein D (2000) Molecular portraits of human breast tumours. Nature 406:747-52

Pomeroy SL, Tamayo P, Gaasenbeek M, Sturla LM, Angelo M, McLaughlin ME, Kim JY, Goumnerova LC, Black PM, Lau C, Allen JC, Zagzag D, Olson JM, Curran T, Wetmore C, Biegel JA, Poggio T, Mukherjee S, Rifkin R, Califano A, Stolovitzky G, Louis DN, Mesirov JP, Lander ES, Golub TR (2002) Prediction of central nervous system embryonal tumour outcome based on gene expression. Nature 415:436-42

Radmacher MD, McShane LM, Simon R (2002) A paradigm for class prediction using gene expression profiles. J Comput Biol 9:505-11

Ramaswamy S, Tamayo P, Rifkin R, Mukherjee S, Yeang CH, Angelo M, Ladd C, Reich M, Latulippe E, Mesirov JP, Poggio T, Gerald W, Loda M, Lander ES, Golub TR (2001) Multiclass cancer diagnosis using tumor gene expression signatures. Proc Natl Acad Sci U S A 98:15149-54

Rifkin R, Mukherjee S, Tamayo P, Ramaswamy S, Yeang CH, Angelo M, Reich M, Poggio T, Lander ES, Golub TR, Mesirov J (2003) An Analytical Method For Multi-class Molecular Cancer Classification. SIAM Review 45:706-723

Rosenwald A, Wright G, Chan WC, Connors JM, Campo E, Fisher RI, Gascoyne RD, et al. (2002) The use of molecular profiling to predict survival after chemotherapy for diffuse large-B-cell lymphoma. N Engl J Med 346:1937-47

Sarwal M, Chua MS, Kambham N, Hsieh SC, Satterwhite T, Masek M, Salvatierra O, Jr. (2003) Molecular heterogeneity in acute renal allograft rejection identified by DNA microarray profiling. N Engl J Med 349:125-38

Shipp MA, Ross KN, Tamayo P, Weng AP, Kutok JL, Aguiar RC, Gaasenbeek M, Angelo M, Reich M, Pinkus GS, Ray TS, Koval MA, Last KW, Norton A, Lister TA, Mesirov J, Neuberg DS, Lander ES, Aster JC, Golub TR (2002) Diffuse large B-cell lymphoma outcome prediction by gene-expression profiling and supervised machine learning. Nat Med 8:68-74

Simon R, Radmacher MD, Dobbin K, McShane LM (2003) Pitfalls in the use of DNA microarray data for diagnostic and prognostic classification. J Natl Cancer Inst 95:14-8

Sorlie T, Perou CM, Tibshirani R, Aas T, Geisler S, Johnsen H, Hastie T, Eisen MB, van de Rijn M, Jeffrey SS, Thorsen T, Quist H, Matese JC, Brown PO, Botstein D, Eystein Lonning P, Borresen-Dale AL (2001) Gene expression patterns of breast carcinomas distinguish tumor subclasses with clinical implications. Proc Natl Acad Sci U S A 98:10869-74

Sorlie T, Tibshirani R, Parker J, Hastie T, Marron JS, Nobel A, Deng S, Johnsen H, Pesich R, Geisler S, Demeter J, Perou CM, Lonning PE, Brown PO, Borresen-Dale AL, Botstein D (2003) Repeated observation of breast tumor subtypes in independent gene expression data sets. Proc Natl Acad Sci U S A 100:8418-23

Tan PK, Downey TJ, Spitznagel EL, Jr., Xu P, Fu D, Dimitrov DS, Lempicki RA, Raaka BM, Cam MC (2003) Evaluation of gene expression measurements from commercial microarray platforms. Nucleic Acids Res 31:5676-84

Tibshirani R, Hastie T, Narasimhan B, Chu G (2002) Diagnosis of multiple cancer types by shrunken centroids of gene expression. Proc Natl Acad Sci U S A 99:6567-72

Tukey JW (1993) Tightening the clinical trial. Control Clin Trials 14:266-85

Tusher VG, Tibshirani R, Chu G (2001) Significance analysis of microarrays applied to the ionizing radiation response. Proc Natl Acad Sci U S A 98:5116-21

van de Vijver MJ, He YD, van't Veer LJ, Dai H, Hart AA, Voskuil DW, Schreiber GJ, Peterse JL, Roberts C, Marton MJ, Parrish M, Atsma D, Witteveen A, Glas A, Delahaye L, van der Velde T, Bartelink H, Rodenhuis S, Rutgers ET, Friend SH, Bernards R (2002) A gene-expression signature as a predictor of survival in breast cancer. N Engl J Med 347:1999-2009

van 't Veer LJ, Dai H, van de Vijver MJ, He YD, Hart AA, Mao M, Peterse HL, van der Kooy K, Marton MJ, Witteveen AT, Schreiber GJ, Kerkhoven RM, Roberts C, Linsley PS, Bernards R, Friend SH (2002) Gene expression profiling predicts clinical outcome of breast cancer. Nature 415:530-6

Vasselli JR, Shih JH, Iyengar SR, Maranchie J, Riss J, Worrell R, Torres-Cabala C, Tabios R, Mariotti A, Stearman R, Merino M, Walther MM, Simon R, Klausner RD, Linehan WM (2003) Predicting survival in patients with metastatic kidney cancer by gene-expression profiling in the primary tumor. Proc Natl Acad Sci U S A 100:6958-63

West M, Blanchette C, Dressman H, Huang E, Ishida S, Spang R, Zuzan H, Olson JA, Jr., Marks JR, Nevins JR (2001) Predicting the clinical status of human breast cancer by using gene expression profiles. Proc Natl Acad Sci U S A 98:11462-7

Wigle DA, Jurisica I, Radulovich N, Pintilie M, Rossant J, Liu N, Lu C, Woodgett J, Seiden I, Johnston M, Keshavjee S, Darling G, Winton T, Breitkreutz BJ, Jorgenson P, Tyers M, Shepherd FA, Tsao MS (2002) Molecular profiling of non-small cell lung cancer and correlation with disease-free survival. Cancer Res 62:3005-8

Yuen T, Wurmbach E, Pfeffer RL, Ebersole BJ, Sealfon SC (2002) Accuracy and calibration of commercial oligonucleotide and custom cDNA microarrays. Nucleic Acids Res 30:e48

Chapter 3

THE NEEDED REPLICATES OF ARRAYS IN MICROARRAY EXPERIMENTS FOR RELIABLE STATISTICAL EVALUATION

Sue-Jane Wang,[1] James J. Chen[2]
[1]Division of Biometrics II, Office of Biostatistics, Office of Pharmacoepidemiology and Statistical Science, Center for Drug Evaluation and Research, [2]Division of Biometry and Risk Assessment, National Centerfor Toxicologic Research, U.S. Food and Drug Administration

Abstract: Microarray technology provides exciting tools for monitoring expression levels of hundreds or thousands of genes simultaneously. Good microarray studies have clear objectives. To make meaningful statistical interpretation of study results obtained from microarray experiments, the design of the experiments must consider some degree of replication to allow for the description of sources of variations. In this article, we present an overview of replicate designs that incorporate measurement variability to address its study objectives.

Key words: Differentially expressed genes; level of significance; power; sample size; standardized effect size; study objective

1. INTRODUCTION

The problem of calculating the number of arrays needed in microarray experiments is similar to the problem of calculating the sample size and power in clinical trials or other scientific experiments, with the caveat that microarray experiments involve hundreds or thousands of genes, only a fraction of which is expected to be differentially expressed or termed altered.

Replicates allow for assessment of variation in expression data so that formal statistical analysis methods can be applied. Without replicates, one cannot distinguish between true differences in gene expression versus random fluctuations. Replicates can take place at different levels of the experiment. For example, replicates can be conducted for different tissues or different cell lines. Each one can be hybridized to more than one array and each array can consist of replicated spots of the same gene. Yang and Speed [2002] described two types of replication: biological replicates and technical replicates. An example of biological replicates is a set of hybridizations that involve mRNA from different extractions. The biological samples are the experimental units. Statistical tests should be based on the biological replicate samples to determine the effects of a treatment on different biological populations. Technical replicates include replicates in which the mRNA is from the same pool (the same extraction). Technical replicates are used to detect variation within the experimental groups. In general, an experimenter will use biological replicates to obtain averages of independent data and to validate a generalization of the conclusion and technical replicates to assist in reducing experimental variability.

In the classical (or frequentist) approach, the sample size estimation for planning a comparative scientific experiment requires specification of two hypotheses (a null hypothesis and an alternative hypothesis), the Type I error (α-error), the Type II error (β-error), the corresponding power level, and a targeted effect size. For instance, the effect size may be an n-fold change of a gene tested under two conditions. Using clinical trials as an example, the sample size is the number of patients needed to conduct a clinical trial aimed at demonstrating a better treatment regimen relative to some standard of care or placebo vehicle. This number is calculated based on a fixed Type I error rate and a pre-specified power level to detect a pre-specified treatment effect size. Table 1 contains a glossary of terms often used in statistical inference and in the design of an experiment in sample size estimation.

Table 1. A glossary of key terms used for statistical inference, sample size estimation, and power analysis for a scientific experiment.

GLOSSARY	
clinical hypothesis	an assumption stated in plain text describing its intended objective(s)
statistical hypothesis	an assertion or conjecture about the probability distribution for the designated population parameter(s) with respect to clinical objective(s)
null hypothesis	the status quo statement about the parameter value under investigation, e.g., the log mean expression intensity of a gene is no different between the treated and the control tissues
alternative hypothesis	any admissible conjecture that does not overlap with the null hypothesis; the study objective of interest is generally designated as the alternative hypothesis, *i.e.*, the claim to be proved

GLOSSARY	
hypothesis test	a procedure driven by a data-based rule for deciding whether to accept the null hypothesis or to reject it in favor of the alternative hypothesis
population	a collection of all individuals (units) of interest
sample	the subset of a population that is actually observed
random sample	a collection of data of experimental units selected from a population. Each member of the population has an equal pre-assigned chance of being selected
parameter	a numerical characteristic of a population
statistic	a summary numerical characteristic calculated from a sample that is used to infer values of parameters
type I error (α-error)	an error made by rejecting a null hypothesis when the null hypothesis is true
α-risk	the probability of making a type I error
type II error (β–error)	an error made by accepting a null hypothesis when the alternative hypothesis is true
β risk	the probability of making a type II error
level of significance	the upper bound on probability of type I error, which is usually a small number, e.g., 0.01, 0.05
p-value	the observed significance level
power of the test	the probability of rejecting the null hypothesis when it is false
test statistic	a summary numerical characteristic calculated from a sample, whose value is used to decide which of two statistical hypotheses should be accepted as true
effect size	the targeted distance between the parameters of two populations designed to detect, e.g., the difference in population means, the ratio in population variances
sample size	the number of experiment units, e.g., arrays or biological replicates in a microarray experiment
multiplicity	multiple hypotheses tested simultaneously in an experiment
familywise error rate	the probability of making at least one false rejection among all hypotheses tested
false discovery rate	the expected proportion of false rejections among the rejected null hypothesis

In comparative experiments, assume that there are two experimental conditions (e.g., normal tissue vs. diseased tissue, placebo group vs. treated group, etc.) under investigation. Among hundreds or thousands of genes simultaneously studied in a microarray experiment, a gene either expresses equally in the two conditions or it expresses differently (up- or down-regulated). The true state of affairs and the decision to accept or reject a null hypothesis can be summarized in a 2x2 table (Table 2).

Table 2. **A microarray experiment to detect the up-regulation or down-regulation of one gene between two experimental conditions.**

		Decision based on one hypothesis test	
		Do not reject "equal expression"	Reject "equal expression"
True State of Affairs	Equal expression between two conditions	(cannot conclude that a gene is up or down regulated) Correct Decision	Type I Error
	Up or down regulated gene	Type II Error	(conclude that a gene is differentially expressed) Correct Decision

Consider a microarray experiment with adequate power to detect altered (up- or down- regulated) genes, if a gene is truly non-differential, the decision will likely be that stated in the upper left box. Conversely, if gene expressions differ between the two conditions, the decision will tend to be that described in the lower right box. These are correct decisions. When the decision being made is inconsistent with the true state of affairs, errors occur usually with small probabilities, e.g., incorrectly concluding a non-differential gene as differential (Type I error) or a differential gene as non-differential (Type II error).

2. TYPICAL SAMPLE SIZE ESTIMATION IN PLANNING A COMPARATIVE EXPERIMENT

The decisions in hypothesis testing rely on a specific statistical approach to formulate the probability function under the two hypotheses in delineating its likelihood function. Below, we illustrate the sample size estimation for a single gene.

2.1 Statistical Models

Denote the background-subtracted normalized intensity (e.g., in log based-2 scale) for control and treated samples by $\boldsymbol{Y_c}$ and $\boldsymbol{Y_t}$, respectively, for the spot (gene) g ($g = 1, ..., G$) and replicate j ($j = 1, ... J$). For simplicity, we drop the subscripts g and j and consider a balanced design. That is, the number of arrays needed in the control and treated sample is the same. When the variances of the two groups are equal, the balanced design is optimal in terms of the total sample size needed at the desired power level. A model for Y_c and Y_t is $Y_c = \mu_c + \varepsilon_c$; $Y_t = \mu_t + \varepsilon_t$, where μ_c and μ_t are true expression levels for control and treated samples, respectively. For each gene g in replicate j, the errors ε_c and ε_t are assumed to be independently and identically bivariate-normally distributed $(\varepsilon_c, \varepsilon_t) \sim N(0, \Sigma)$, where

$$\Sigma = \begin{bmatrix} \sigma_c^{\;2} & \rho\sigma_c\sigma_t \\ \rho\sigma_c\rho_t & \sigma_t^{\;2} \end{bmatrix}$$

The correlation ρ can be non-zero for the two-color fluorescence experiment. The random variable $T = Y_c - Y_t$ is normally distributed with mean $(\mu_c - \mu_t)$ and variance $\sigma^2 = \sigma_c^{\;2} - 2\rho\sigma_c\sigma_t + \sigma_t^{\;2}$.

Identifying differentially expressed genes between the control and the treatment can be formulated in terms of the hypotheses

$$H_{0,g}: \mu_c - \mu_t = 0 \quad \text{versus} \quad H_{1,g}: \mu_c - \mu_t \neq 0,$$

for $g = 1, ..., G$. The unstandardized sampling statistic $\bar{Y}_c - \bar{Y}_t$ is used to test the hypothesis $H_{0,g}$, where $\bar{Y}_c$ and $\bar{Y}_t$ are the means of the J replicates in the control group and J replicates in the treatment group, respectively. The hypothesis test is commonly done by computing the two-sample (standardized) t-statistic $(\bar{Y}_c - \bar{Y}_t)/\hat{\sigma}_2$, where $\hat{\sigma}_2$ is the standard error estimate of $(\bar{Y}_c - \bar{Y}_t)$. The standard error estimate can be computed in two ways, one is to assume that the population standard deviation from the two groups are the same, $\hat{\sigma}_2 = s_p\sqrt{(2/J)}$, where $S_p = \sqrt{(s^2_{\;1} + s_2^{\;2})/2}$, and $s^2_{\;1}$ and $s^2_{\;2}$ are the sample variances for the respective samples, and the standard error estimate is simplified to $\hat{\sigma}_2 = \sqrt{(s^2_{\;1} + s^2_{\;2})/J}$. The other is that the true standard deviations are assumed to be different. In general, if the sample sizes are different between the two groups, say, n_1 for the treated group and n_2 for the control group, then, the standard error estimate becomes $\hat{\sigma}_2 = \sqrt{(s^2{}_1/n_1 + s^2{}_2/n_2)/(n_1 + n_2 - 2)}$. Under the model of an equal variance $\mathrm{var}(Y_c) = \mathrm{var}\,(Y_t)$, if there is no difference between the two groups, then $(\bar{Y}_c - \bar{Y}_t)/\hat{\sigma}_2$ has a t-distribution with $2J - 2$ degrees of freedom. For unequal variances, an approximate t-distribution with degree of freedom v can be computed, known as the Welch-Satterthwaite method, $v = (w_1 + w_2)^2/(w_1^{\;2}/(n_1 - 1) + w^2{}_2/(n_2 - 1))$, where $w_1 = s^2{}_1/n_1, w_2 = s^2{}_2/n_2$.

When the number of replicates is the same for the two groups, the degrees of freedom become $v = (J-1)(s^2_1 + s^2_2)/(s^4_1 + s^4_2)$. The comparison between the control and treatment groups can also be tested by a one-sample t-statistic. Let T be the mean of the $T_1, \ldots, T_J$, where $T = Y_c - Y_t$ as defined previously. If there is no difference between the two groups, then the one-sample standardized statistic $(\bar{Y}_c - \bar{Y}_t)/\hat{\sigma}_1$ has a t-distribution with $J - 1$ degrees of freedom, where $\hat{\sigma}_1$ is the standard error estimate of T. Under the assumption of an equal variance in the two groups, the variance is $\hat{\sigma}_1^2 = [2(1-\rho)/J]\sigma^2$. One-sample t-test is a more powerful test if the correlation $\rho > 0$.

2.2 Sample Size Calculation

For a given significance level α, the power of the two-sample t-test for gene g is $\gamma = F[\sqrt{J/2}\Delta_2 - t_{\alpha/2}]$, where $\Delta_2 = (\mu_c - \mu_t)/\sigma_2$ and F is the cumulative t_{2J-2} distribution (Student's t-distribution with $2J - 2$ degrees of freedom) and $t_{\alpha/2}$ is the $100(\alpha/2)th$ percentile of the t_{2J-2} distribution. For specified α, γ, and Δ_2, the sample size needed in each group to detect a significance for gene g is

$$J = 2(t_{\alpha/2} - t_{\gamma})^2 / \Delta_2^2 \tag{1}$$

The power and the sample size given above are for the gth gene with a specified standardized effect size Δ_2, $g = 1, \ldots, G$ for the two-sample t-test approach. They can be similarly obtained for the one-sample t-test. The power for the one-sample t-test for any gene g is $\gamma = F[\sqrt{J}\Delta_1 - t_{\alpha/2}]$, where F *is* the cumulative t-distribution with $(J - 1)$ degrees of freedom and Δ_1 is the standardized effect size for one-sample approach. The corresponding sample size needed to detect significance for a gene g is

$$J = (t_{\alpha/2} - t_{\gamma})^2 / \Delta_1^2 \tag{2}$$

Instead of t-distribution, a standard normal distribution, also called z-distribution, is often applied to Equations (1) and (2) for the two-sample approach and the one-sample approach, respectively, by assuming that the true variability of the distribution of interest can be reasonably approximated. It is imperative to note that the exact quantities $t_{\alpha/2}$ and t_{γ} with their corresponding degrees of freedom in Equations (1) and (2) are obtained using iterative simulation procedure, *viz.*, the concept described in the Appendix of Elston et al. [1999].

3. STATISTICAL APPROACHES TO NEEDED REPLICATES IN MICROARRAY EXPERIMENTS

The experimental design and statistical analysis for a microarray experiment should reflect the overall objectives of the study. A common objective is the identification of differentially expressed genes, e.g., investigating differences in gene expression profiles from different tissue types or from the same tissue with and without exposure to a specific drug or toxicant. Another common objective is to develop multi-gene predictors of class for a sample using its gene expression profile. In both class comparison and class prediction objectives, the classes being compared or used to predict the class membership on the basis of influential gene expressions are predefined. Other examples of common objectives include studying relationships between genes and gene clusters or relationships between genes and clinical predictors or outcomes, and studying changes in expression profiles over time or dosage, etc. Most of the statistical methods in the literature for estimating array replicates center on the objective of identification of differentially expressed genes.

3.1 Multiple Testing Context

As microarray studies typically monitor expression levels of thousands of genes simultaneously, the probabilities of making incorrect test conclusions, false positives and false negatives should be considered. Table 3 is an extension of Table 2 involving the entire gene set tested.

Type I error probability is one of the most important error measures in statistical significance testing. Type I error conventionally refers to rejection of the true null hypothesis. There are many possible Type I error measures under multiple hypotheses testing. The expected number of false positives is $E(V)$. The expected proportion of false positives among the G tests is $E(V)/G$, the per comparison error rate (*CWE*). When $G = 1$, the *CWE* is the α risk. Since hundreds or thousands of tests are conducted, simply using the *CWE* significance level without adjusting for multiple tests will increase the chance of false positive findings. The traditional approach is to control the probability of rejecting at least one true null hypothesis, the familywise error rate, $FWE = Pr(V > 0)$ [Hochberg and Tamahane, 1987; Westfall and Young, 1993]. That is, the *FWE* approach guarantees that the probability of one or more false positives is not greater than a pre-determined level, regardless of how many genes are tested. When the number of genes is very large such as microarray data, the use of *FWE* criterion may require a large number of samples.

Table 3. **A microarray experiment to detect genes that are differentially expressed between two conditions in multiple testing framework.**

		Test declaration		
		Reject "equal expression"	Do not reject "equal expression"	Number of genes
True State of Affairs	Equal expression between two conditions	V	S	G_0
	Up or down regulated gene	U	T	G_1
	Total	R	A	G

R: Total number of genes the test concludes altered (H_0 rejected)
V: Number of false positives declared by the test (type I errors)
U: Number of true positives and the test declares as differentially expressed
A: Total number of genes that the test concludes not differentially expressed between the two conditions (H_0 accepted)
S: Number of true negatives and the test declares as not differentially expressed
T: Number of false negatives declared by the test (type II errors)
G: Total number of genes tested
G_0: Number of genes that are truly not differentially expressed
G_1: Number of genes that are truly differentially expressed

Benjamini and Hochberg [1995] proposed the false discovery rate *(FDR)* as an alternative error measure. *FDR* is the expected proportion of the null hypotheses that are falsely rejected *E(V/R),* if $R > 0$. If all null hypotheses are true ($G_0 = G$), the *FDR* is equivalent to the *FWE.* When $G_0 < G$, the *FDR* is smaller than or equal to the *FWE.* This implies that if a procedure controls the *FWE,* then it controls the *FDR.* Thus, the *FDR* is a less stringent criterion than the *FWE,* therefore it leads to an increase in the power of identifying differentially expressed genes. However, the exact number of true null hypotheses is usually unknown, so the number of errors amongst the rejected hypotheses is also unknown; the use of the *FWE*-controlled approach in sample size calculation will ensure controlling either the *FDR* or *FWE.*

Sensitivity and specificity are two commonly used measures for evaluating the accuracy of a diagnostic test for a disease marker. Sensitivity of a test is defined as the probability that the test is positive, given a disease is present. Specificity is the probability that the test is negative, given a disease is absent. In the context of Table 3, the sensitivity is U/G_I and the specificity is S/G_0. Wang and Chen [2004] formulated a sample size calculation method for microarray experiments by building in the sensitivity parameter for the overall power in addition to individual gene power. Zien et al. [2002] presented a sample size estimation procedure using a mathematical model, which incorporates variability parameters that capture additive and multiplicative measurement errors, and biological variability. Through simulation, they showed a few combinations on the number of replicates, signal-to-noise ratio, and fold ratio of expression between two classes, and their impact to the sensitivity and specificity. Lee and Whitmore [2002] tackled the power as the expected proportion of truly expressed genes that are correctly declared as expressed = $E(U)/G_I$.

3.2 Sample Size Calculation

For specified α, γ, and Δ (Δ_1 or Δ_2), the sample sizes needed in each group to detect significance for gene *g* are given in Equations (1) and (2). In practice, only a fraction of the genes will be affected by a treatment in the experiment, *viz.*, genes that are differentially expressed between the two groups. Wang and Chen [2004] formulated the sample size problem as: the number of arrays needed to detect at least $100\lambda\%$ of the truly differentially expressed genes at the desired overall power $(1-\beta)$, where λ is a pre-specified fraction, $0 < \lambda \leq 1$. For illustrative purpose, let's assume an equal effect size for all altered genes, and the effect size is standardized by its standard deviation. This assumption can be relaxed in application. For a specified significance level α, a standardized effect size Δ (the desired change of a gene to detect between the treatment and control samples divided by the estimated standard deviation), and *J* replicates per group, the power of detecting a truly altered gene is given in Equation (3). Let *k* (*i.e.*, G_I in Table 3) denote the number of truly differentially expressed genes, and the notation $[t] = b$ denote the largest integer less than *t*. The power for identifying at least $[k\lambda+1] = b$ altered genes can be computed by summing the binomial probabilities:

$$1-\beta=\sum_{l=b}^{k}(k!/(l!(k-l)!))\ \gamma^{l}(1-\gamma)^{k-l} \tag{3}$$

Recall that γ is the power level for any gene g described in Section 2. Given k, λ, and β, the γ can be estimated by solving the above equation. Thus, the power and the sample size for the two-sample t-test or the one-sample t-test can be obtained. Here, the sample size is calculated so as to achieve the objective of identifying at least $100\lambda\%$ of truly altered genes at the overall power $1\text{-}\beta$. It is important to note that Equations (1) and (2) are now utilized to compute the replicate numbers needed, in which γ may not be the usual individual power level, such as, 80% or 90% level, but, the power level for a given gene g required in order to reach the overall power level, say, 80% or 90%. For the two-sample z-test or the one-sample z-test, the sample sizes are to be estimated by substituting the t distribution function by the z distribution function in Equations (1) and (2), respectively.

Clearly, the sample size depends on the choice of significance level α. The selection of an overall α-level will have impact on the sample size calculation when more than one gene is differentially expressed. There are two underlying approaches to the choice of α, the *FWE*-controlled Bonferroni approach using $\alpha = 0.05/G$ and the unadjusted $CWE = \alpha$. The Bonferroni approach is known to lack power when the number of genes is large or the genes are correlated. An *FWE* approach can be improved by incorporating the dependent structure among genes, such as the resampling techniques of Westfall and Young [1993], in the analysis. However, the dependency of data structure is not available at the time of sample size planning. The Bonferroni approach guarantees the controlling of the *FWE*, regardless of the true correlation structure, which is usually unknown. The Bonferroni approach explicitly accounts for the number of genes tested. With the unadjusted *CWE* approach, though it ignores the multiple testing, the α level can be chosen to reflect experimental objectives. Investigators might be more interested in identifying potential genes that are differentially expressed with a limited number of false positive findings. For example, using $\alpha = 0.001$ will result in false positives of 1 per 1,000 non-differentially expressed genes. This paper takes the same principle of fixing a rejection level α for individual hypotheses beforehand; the choice of α can be based on either the Bonferroni or the *CWE* approach depending on the objective of the experiment.

3.3 Numerical Results

Two practical scenarios to illustrate sample size calculation are presented. The first scenario considers an initial exploration intent on analyzing ten thousand genes for the purpose of examining as many interesting genes as possible, treating the experiment as a screening tool. The second scenario considers a more focused exploration on (selected) one thousand genes. The parameters used for Scenario 1 are as follows. Suppose $100\eta\%$ out of the 10,000 genes studied are truly differentially expressed. To account for a large number of tests (genes), we did the analysis for two significance levels: (a) the Bonferroni adjustment with an overall significance level *(FWE)* at 0.05 ($\alpha = 0.05/10{,}000$ for any individual gene) and (b) a constant comparison-wise error rate *(CWE)* set at 0.001 regardless of the number of hypotheses tested. At the given significance level, we target an overall power of 80% to detect at least $\lambda = 0.5, 0.7, 0.9$ and 1.0 of the altered genes for effect sizes $\Delta = 2$ and 4.

The numbers of replicated arrays needed are tabulated in Table 4 for $\eta =$ 0.05, 0.10 and 0.20 using *t*-test approach versus using *z*-test approach. As an example, for $\lambda = 0.9$, $\eta = 0.05$, and $\Delta = 2.0$ (to detect at least 450 genes of the 500 truly altered genes, with the effect size 2, from 10,000 genes studied), using the one-sample *t*-test with *CWE* $\alpha = 0.001$, the number of arrays needed in each group is 10. This number is six using the *z*-test approach. For scenario 2, only 1000 genes are studied. The proportions of truly altered genes considered are $\eta = 0.1, 0.2, 0.5$ and 0.8. The remaining parameters are the same as those in Scenario 1. The patterns observed for Scenario 1 and Scenario 2 are similar (results not shown). Results tabulated in Table 4 were validated via Monte Carlo simulations assuming the log-intensity data are normally distributed.

To illustrate how the number of replicates can be planned for an experiment aiming for identification of differentially expressed genes, a pilot study based on a cDNA experiment for a toxicogenomic study of gene expression levels of kidney samples from rats dosed with a drug was used to estimate the standardized effect size.

Briefly, the experiment included six arrays from the 700 gene rat Phase-1 chip. In addition, sequences of five genes from other species different from the one of 700 genes were also spotted on the array to monitor non-specific background binding of labeled mRNA serving as the housekeeping genes. The normalized data using the approach of Chen et al. [2002] was adopted to estimate the standardized effect size.

Consider the two-sample approach applicable to a reference design of a microarray experiment. (In the reference design, all samples of interest (control and treatments) are hybridized on different arrays labeled with the same color dye, while a reference sample labeled with the other color dye is used on every array to hybridize with either a control or a treatment sample.) Using the example data set, the estimated standardized effect sizes are -1.7 using the 95*th* percentile for the down-regulated genes and 2.1 using the 5*th* percentile for the up-regulated genes. One could consider an estimated absolute standardized effect size of 2. If the number of genes to be studied is 10,000, then one can use Table 4 to estimate the number of arrays. Using the

CWE criterion as an example, to identify at least 90% of truly altered genes for the two-sample case, 14 arrays are needed. If dye-swap in a two-color system is designed, the one-sample approach may be adopted. One may consider an estimated absolute standardized effect size of 6. With the same criteria as above, the estimated number of arrays would reduce to five.

Table 4. The number of replicates needed to detect at least λ = 0.5, 0.7, 0.9 and 1.0 truly altered genes with standardized effect sizes Δ = 2 and 4, and desired power $1\text{-}\beta = 0.8$.

		Two-sample t-test**						One-sample t-test					
		Bonferroni			CWE			Bonferroni			CWE		
Δ	λ	5%*	10%	20%	5%	10%	20%	5%	10%	20%	5%	10%	20%
2.0	0.5	17	17	16	9	9	9	14	14	14	8	8	8
	0.7	19	19	19	11	11	11	15	15	15	9	9	9
	0.9	22	22	22	14	14	14	17	17	17	10	10	10
	1.0	36	38	39	25	27	28	24	25	26	17	18	19
4.0	0.5	8	8	8	5	5	5	9	9	9	5	5	5
	0.7	9	9	9	6	6	6	9	9	9	6	6	6
	0.9	10	10	10	6	6	6	10	10	10	6	6	6
	1.0	13	13	14	9	10	10	12	12	12	8	9	9
		Two-sample z-test**						One-sample z-test					
		Bonferroni			CWE			Bonferroni			CWE		
Δ	λ	5%*	10%	20%	5%	10%	20%	5%	10%	20%	5%	10%	20%
2.0	0.5	12	12	12	7	7	6	6	6	6	4	4	4
	0.7	14	14	14	8	8	8	8	8	8	5	5	5
	0.9	18	18	18	12	12	12	10	10	10	6	6	6
	1.0	32	34	35	23	24	25	17	17	18	12	13	13
4.0	0.5	4	4	4	2	2	2	2	2	2	2	2	2
	0.7	4	4	4	3	3	3	3	3	3	2	3	2
	0.9	5	5	5	4	4	4	3	3	3	2	2	2
	1.0	9	9	10	6	7	7	5	5	5	4	4	4

** Bonferroni: $\alpha = 0.05/10000$, CWE: $\alpha = 0.001$.

* Assuming 5% (500 genes), 10% (1000 genes), and 20% (2000 genes) truly altered genes among the 10,000 genes and λ is the desired fraction of truly altered genes to identify. Ten thousand genes are studied.

In application, the assumed equal standardized effect size can be replaced by the minimum, mean or some percentile of the standardized effect size among all genes. When it is the minimum, the approach gives conservative estimated number of replicates. Depending on the study objective, the standardized effect size can also be the expected log intensity ratio divided by a pre-specified variability or by maximum variability resulting in conservative sample size estimation.

4. DISCUSSION

In clinical trials, sample size estimation often uses z-test approach assuming a known targeted variability learned from earlier clinical experience. In microarray experiments, however, the variability of gene expression intensity measurements is difficult to assume and is usually estimated from limited preliminary experiments. Thus, the *t*-test approach is generally recommended for estimating the needed replicates in microarray experiments. Using the same criteria and parameter estimates, the number of replicates is less with the one-sample approach in a dye-swap two-color design than with the two-sample approach in a reference design.

By and large, two general approaches are utilized in sample size calculation: a parametric modeling approach and a non-parametric approach. The sample size formulae presented above are derived from the parametric normal model and is applicable primarily for biological replicates with no gene replicates within an array. Tsai et al. [2003] showed in a Monte Carlo simulation that when the number of replicates is eight or more, the Type I errors and powers of the parametric *t*-test and permutation test are very similar for normally distributed log-intensity data. When the number of replicates is small, e.g., three or less, the power with *t*-test is very low. Black and Doerge [2002] proposed a parametric approach using the log-normal ANOVA model to estimate the minimum number of spots needed within an array. Pan et al. [2002] proposed a nonparametric normal mixture model approach to calculate the number of replicates required to detecting changes in gene expression. Because of the specific statistic (*t*-type test statistic defined as difference in the means represented in unit of standard deviations), their approach assumed the number of arrays needed is an even integer. From a mixture model analysis, Lee et al. [2000] suggested that three replicates might be sufficient in view of large nose-to-signal ratio. In general, if the distribution is not normally distributed, then the needed sample size can be estimated by a simulation method [Wang and Chen, 2004)]

Ideker et al. [2000] Herwig et al. [2001], Simon et al. [2002] offered advice on the number of array replicates of the same biological sample required in order to reliably identify differences. It is noted that no method is unanimously optimal for all kinds of data. Thus, selection of a method for sample size estimation should be subject to the intended study objectives, the characteristics of the data including the anticipated sources of variations, and the extent of violation of the underlying assumptions.

A few authors proposed methods for replicate estimation to save costs. Their intents are not necessarily to increase the number of arrays, rather, to make use of the existing arrays more efficiently. Donovan and Becker [2002] considered an experimental approach, termed double-round hybridization of membrane based cDNA arrays, to providing replicate data sets without the need of additional arrays and additional probe labeling. This adds little extra cost. Essentially, their ideas to use double round hybridization are to rescue the lost experiment if it occurred, improved background reduction, and may help produce reliable replicate data when the first round hybridization is successful or when a precautionary first round hybridization is performed with a blank nylon membrane.

When the replicates planning hinge on the trade-off of costs between the number of experimental subjects versus the number of arrays, Cui and Churchill [2003] provided a formula to compute the optimum number of arrays per mouse so as to minimize the total cost of the experiment. Chen et al. [2004] gave guidance on optimizing the number of subjects and on between and within array replicates. CAMDA 2003 provides four lung cancer datasets. Many authors investigated the prognostic ability of the gene expressions to the survival of lung cancer patients. Jung et al. [2004] (see Chapter 8 of this book) presented power analyses by setting the first D genes to be prognostic with some correlation, say, *r*, with log survival time and suggested that this inferential method should serve as an helpful tool for sample size and power calculations in designing microarray experiments for which association to survival endpoints are to be studied.

In the case of concerns with tissue sample availability to study transcriptional profiling using microarray and its connection to disease classification, Hwang et al. [2002] proposed a method to determine the minimum array replicates based on the linear combinations of individual genes as variables in the disease classifier using Fisher discriminant analysis, where the individual genes were first selected to be differentially expressed across disease subtypes, the so-called discriminatory genes, using Wilks' lamda score and leave one out cross-validation. Essentially, they estimated the sample size using the combination of a much smaller number of genes (dimensions) involved and allowed refinement of sample size estimation by looping through the process using updated estimated effect sizes and correlations, an adaptive sample size estimation approach.

5. SUMMARY

Variability in microarray data is expected and unavoidable. Replication is the key to the accuracy and reliability of the data. Replication enables us to understand and interpret the significance of observed changes for thousands of genes. It is noteworthy to point out that replication is not to duplicate the results; rather it is to understand the sources of noise so as to control it or reduce it for performing sensible statistical analyses and for drawing reliable inferences.

REFERENCES

Benjamini, Y., Hochberg, Y., 1995, Controlling the false discovery rate: a practical and powerful approach to multiple testing, *Journal of Royal Statistical Society B* **57**: 289-300.

Black, M.A., Doerge, R.W., 2002, Calculation of the minimum number of replicate spots required for detection of significant gene expression fold changes for cDNA microarrays. Technical Report, Department of Statistics, Purdue University, IN.

Chen, Y-J., Kodell, R.L., Sistare, F., Thompson, K., Morris, S., Chen, J.J., 2002, Normalization methods for cDNA microarray data analysis, *Journal of Biopharmaceutical Statistics* **13**: 56-74.

Chen, J.J., Delongchamp, R.R., Tsai, C., Hsueh, H-M., Thompson, K.L., Desai, V.G., Fuscoe, J.C., 2004, Analysis of variance components in gene expression data, *Bioinformatics* **20**, in press.

Cui, X., Churchill, G.A., 2003, How many mice and how many arrays? Replication in mouse cDNA microarray experiments, Johnson KF, Lin SM. *Methods of Microarray Data Analysis III,* Kluwer Academic Publishers, 139-154.

Donovan, D.M., Becker, K.G., 2002, Double round hybridization of membrane based cDNA arrays: improved background reduction and data replication. *Journal of Neuroscience Methods* **118**:59-62.

Elston, R.C., Idury, R.M., Cardon, L.R., Lichter, J.B., 1999, The study of candidate gene in drug trials: sample size considerations, *Statistics in Medicine* **18**:741-751.

Herwig, R., Aanstad, P., Clark, M., Lehrach, H., 2001, Statistical evaluation of differential expression on cDNA nylon arrays with replicated experiments, *Nucleic Acids Research* **29**(23): e117.

Hochberg, Y., Tamahane, A.C., 1987, *Multiple Comparison Procedures,* John Wiley Sons: NY.

Hwang, D., Schmitt, W.A., Stephanopoulos, G., 2002, Determination of minimum sample size and discriminatory expression patterns in microarray data, *Bioinformatics* **18**(9): 1184-93.

Ideker, T., Thorsson, V., Siegel, A.F., Hood, L., 2000, Testing for differentially-expressed genes by maximum-likelihood analysis of microarray data, *Journal of Computational Biology* **7**(6):805-817.

Jung, S-H., Owzar, K., George, S. 2004, Associating microarray data with a survival endpoint, Shoemaker, J, Lin SM, eds, *Methods of Microarray Data Analysis IV,* Kluwer Academic publishers, 109-120.

Lee, M.L.T., Kuo, F.C., Whitmore, G.A., Sklar, J., 2000, Importance of replication in microarray gene expression studies: statistical methods and evidence from repetitive cDNA hybridizations, *Proceeding of National Academy Science USA* **97**:9834-9839.

Lee, M.L.T., Whitmore, G.A., 2002, Power and sample size for DNA microarray studies, Statistics in Medicine **21**:3543-3570.

Pan, W., Lin, J., Le, C.T., 2002, How many replicates of arrays are required to detect gene expression changes in microarray experiments? A mixture model approach Research, *Genome Biology* **3**(5):0022.1-0022.10.

Simon, R., Radmacher, M.D., Dobbin, K., 2002, Design of studies using DNA microarrays, Genetic Epidemiology **23**:21-36.

Tsai, C.A., Chen, Y.J., Chen, J.J., 2003, Testing for Differentially Expressed Genes with Microarray Data, Nucleic Acids Research **31**(9), e52.

Wang, S.J., Chen, J.J., 2004, Sample size for identifying differentially expressed genes in microarray experiments, *Journal of Computational Biology,* in press.

Westfall, P.H., Young, S.S., 1993, *Resampling-Based Multiple Testing,* John Wiley Sons: NY.

Yang, Y.H., Speed, T., 2002, Design issues for cDNA microarray experiments, *Nature Reviews Genetics* **3**:579-587.

Zien, A., Fluck, J., Zimmer, R., Lengauer, T., 2003, Microarrays: How many do you need? *Journal of Computational Biology* **10** (3-4):653-667.

Chapter 4

POOLING INFORMATION ACROSS DIFFERENT STUDIES AND OLIGONUCLEOTIDE CHIP TYPES TO IDENTIFY PROGNOSTIC GENES FOR LUNG CANCER

Jeffrey S. Morris, Guosheng Yin, Keith Baggerly, Chunlei Wu, and Li Zhang
The University of Texas, MD Anderson Cancer Center, 1515 Holcombe Blvd, Box 447, Houston, TX, 77030-4009

Abstract: Our goal in this work was to pool information across microarray studies conducted at different institutions using two different versions of Affymetrix chips to identify genes whose expression levels offer information on lung cancer patients' survival above and beyond the information provided by readily available clinical covariates. We combined information across chip types by identifying "matching probes" present on both chips, and then assembling them into new probesets based on Unigene clusters. This method yielded comparable expression level quantifications across chips without sacrificing much precision or significantly altering the relative ordering of the samples. We fit a series of multivariable Cox models containing clinical covariates and genes and identified 26 genes that provided information on survival after adjusting for the clinical covariates, while controlling the false discovery rate at 0.20 using the Beta-Uniform mixture method. Many of these genes appeared to be biologically interesting and worthy of future investigation. Only one gene in our list has been mentioned in previously published analyses of these data. It appears that the increased statistical power provided by the pooling was key in finding these new genes, since only nine out of the 26 genes were detected when we apply these methods to the two data sets separately, i.e., without pooling.

Key words: Cox regression; meta-analysis; NSCLC; oligonucleotide microarrays

1. INTRODUCTION

The challenge of this CAMDA competition was to pool information across studies to yield new biological insights, improving medical care and leading to a better understanding of lung cancer biology. We selected adenocarcinoma, since most of the available data are from this type of histology, and it is most prevalent in the general population; we decided to focus on the survival outcome. We chose to focus our efforts on the Michigan and Harvard studies. Both studies used Affymetrix oligonucleotide arrays, but they used different versions of Affymetrix chips: the Michigan study used the HuGeneFL while Harvard used the U95Av2.

Our first goal in this work was to pool the data across different studies to identify prognostic genes for lung adenocarcinoma. By prognostic genes, we meant those whose expression levels offer information on patient survival *over and above* the information already provided by known clinical predictors. We predicted that by pooling the data as opposed to merely pooling the results, we would have more statistical power to detect prognostic genes. Accomplishing this goal required us to develop methodology to pool information across different versions of Affymetrix chips in such a way that we obtained comparable expression levels across the different chip types.

2. ANALYTICAL METHODS

2.1 Pooling Information across Studies

Before pooling the studies, we checked to see if they had comparable patient populations, and we found comparable distributions of age, gender, smoking status, and follow-up time in the studies ($p>0.05$ for all). The stage distributions were slightly different, since the Michigan study contained only stage I and stage III cancers (67 and 19, respectively), while the Harvard study contained patients at all four disease stages (76, 23, 11, and 15, respectively). However, the proportions of advanced (stage III and IV) versus local (stage I and II) disease were similar in the two groups (0.22 vs. 0.78 for Michigan, 0.21 vs. 0.79 for Harvard, $p>0.05$). In spite of these similar characteristics, the patients in these two studies demonstrated significantly different survival distributions, with the Harvard patients

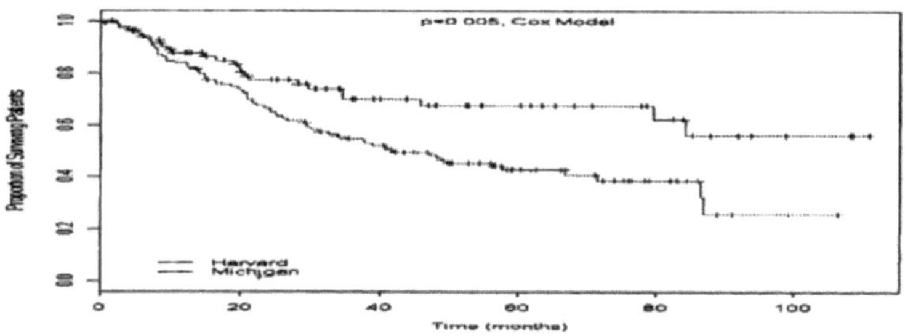

Figure 1. Kaplan-Meier plots for Harvard and Michigan Studies. The p-value corresponds to the institution factor in a multivariable Cox model which also includes age and stage of disease (local/advanced).

tending to have worse prognoses. Figure 1 contains the Kaplan-Meier plots for these two groups. This difference was statistically significant (p=0.005, Cox model) even after adjusting for age and stage, so we included a fixed institution effect in all subsequent survival modeling to account for apparent differences in the patient populations for these two studies.

2.2 Pooling Information across Different Oligonucleotide Arrays using "Partial Probesets"

A major challenge in pooling these studies was that different versions of the Affymetrix Oligonucleotide chip were used in the microarray analyses. The Michigan study used the HuGeneFL Affymetrix chip. This chip contains 6,633 probesets, each with 20 probe pairs. By contrast, the Harvard study used the newer U95Av2 chip. This chip contains 12,625 probesets, each with 16 probe pairs. This difference in chip types raised two problems. First, some genes were represented on one chip but not the other. Second, genes present on both chips were represented by different sets of probes on the two chips. Since the two chip types did not contain the same probesets, we did not expect standard analyses on these Affymetrix-determined probesets to yield comparable expression level quantifications across chips. However, there are some probes that both chips share in common, which we call "matching probes". These probes share common chemical properties on the two chips, and so should yield comparable intensities across the two chip types. Our method focused on these matching probes.

Our first step was to identify the matching probes present on both the HuGeneFL and U95Av2 chips. We next recombined these probes into new probesets using the current annotation of U95Av2 based on Unigene build 160. We refer to these recombined probesets as "partial probesets". Note that because they are explicitly based on Unigene clusters, these probesets will not precisely correspond to the Affymetrix-determined probesets. Frequently, multiple Affymetrix probesets map to the same Unigene cluster. We then eliminated any probesets consisting of just one or two probes, because we expected the summaries from these probesets to be less precise. This left us with 4,101 partial probesets. Most of the probesets (84%) of the probesets contained 10 or fewer probes and the median probeset size was seven. We had several probesets that contained more than 20 probes.

2.3 Preprocessing and Quantifying Gene Expression Levels

We converted the raw intensities for each microarray image to the log scale and re-plotted them to check for poor-quality arrays. We removed from consideration several arrays that have apparent quality problems. From the Michigan data set, samples L54, L88, L89, and L90 contained a large dead spot at the center of the chip, which was obvious when looking at our log-scale plot, shown in Figure 2. These dead spots may have been bubbles caused by inadequate hybridization from using less than the specified 200ul of hybridization fluid. Samples L22, L30, L99, L81, L100, and L102 contained a large number of extremely bright outliers according to MAS5.0. For the Harvard data set, two outlier chips were detected using dChip (CL2001040304 and CL2001041716) and removed. For the Harvard samples with replicate arrays, we kept only the most recently run chip. The remaining data was matching clinical and microarray data for 200 patients, 124 from the Harvard study and 76 from the Michigan study.

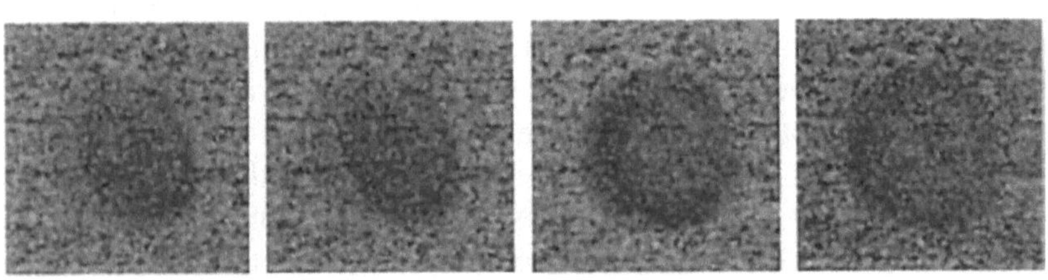

Figure 2. Log intensity plot for four Michigan samples (L54, L88, L89, and L90, respectively) with inadequate hybridization in the middle of the chips.

For each patient, we obtained log-scale quantifications of the gene expression levels for each partial probeset using the Positional Dependent Nearest Neighbor (PDNN) model. This method was introduced in last year's CAMDA competition [Zhang, Coombes, and Xia, 2003], and uses probe sequence information to predict patterns of specific and nonspecific hybridization intensities. By explicitly using the sequencing information, this model was able to borrow strength across probe sets while doing the quantification. This method has been shown to be more accurate and reliable than MAS 5.0 (Affymetrix, Inc.) or dChip [Schadt, et al. , 2001], using the Latin-square test data set provided by Affymetrix for calibrating MAS 5.0 [Zhang, et al., 2003].

We also performed other preprocessing steps. We removed the half of the probesets with the lowest mean expression levels across all samples, then normalized the log expression values by using a linear transformation to force each chip to have a common mean and standard deviation across genes. We next removed the probesets with the smallest variability across chips (standard deviation <0.20), since we considered them unlikely to be discriminatory and more likely to be spuriously flagged as prognostic. Finally, we removed the probesets with poor relative agreement (<0.90) between the partial probeset and full probeset quantifications (see Section 3). After this preprocessing, 1036 probesets remained and were considered in our subsequent analyses.

2.4 Identifying Prognostic Genes

Our main goal was to identify prognostic genes offering predictive information on patient survival. We were interested not primarily in finding genes that were simply surrogates for known clinical prognostic factors like stage, since these factors are easily available without collecting microarray data. Rather, we were interested in finding genes that explain the variability in patient survival that remains after modeling the clinical predictors. Thus, we fit multivariable survival models, including clinical covariates in all survival models we used to identify prognostic genes.

We applied Cox regression models to the survival data combined across both institutions. Our best clinical model included age and disease stage (dichotomized as low, stages I-II, and high, stages III-IV). Smoking status was only marginally significant for survival; therefore, we removed it from the model. Thus, we screened the 1036 genes to find potentially prognostic ones by fitting a series of multivariable Cox models containing age, stage, institution, and the log-expression of one of the genes as predictors. We obtained the exact p-values for each gene's coefficient using a permutation approach. In this approach, we first generated 100,000 datasets by randomly permuting the gene expression values across samples while keeping the clinical covariates fixed. Subsequently, we obtained the permutation p-value for each gene by counting the proportion of fitted Cox coefficients that were more extreme than the coefficient for the true dataset. We also obtained p-values using asymptotic likelihood ratio tests (LRT) and the bootstrap to assess robustness of our results. The results were generally concordant; see Section 4. A small p-value for a given gene indicated potential for that gene to provide prognostic information on survival beyond the clinical covariates.

If there were no prognostic genes, statistical theory suggests that a histogram of these p-values should follow a uniform distribution. An overabundance of small p-values would indicate the presence of prognostic genes. We fit a Beta-Uniform mixture model to this histogram of p-values using a method called the Beta-Uniform Mixture method (BUM, Pounds and Morris, 2003), which partitions the histogram into two components, a Beta component containing the prognostic genes and Uniform component containing the non-significant ones. Various criteria can be used along with this method to determine a cutpoint between these components. We used the false discovery rate (FDR, Benjamini and Hochberg, 1995), which estimates the proportion of genes flagged as prognostic that are in fact not prognostic. Given a choice for FDR, the BUM method yields a p-value cutoff below which a gene is flagged as significant.

We also identified genes differentially expressed by cancer stage by applying the BUM model to p-values from nonparametric Wilcoxon tests comparing median expression levels for early- (stages I-II) and late-stage (stages III-IV) lung adenocarcinoma.

3. ASSESSING "PARTIAL PROBESET" METHOD

Before analyzing the microarray data to identify prognostic genes, we assessed whether our method for combining information across different Affymetrix chip types performed acceptably. First, we checked whether the expression levels were indeed comparable across chip types. Figure 3 contains plots of the median and median absolute deviation (MAD) log expression level for each partial probeset across the Michigan samples run on the HuGeneFL chip against those from the Harvard samples run on the U95Av2 chip. The concordance between these values was 0.961 for the median and 0.820 for the MAD, so it appears that our method yielded reasonably comparable expression levels across the two chips.

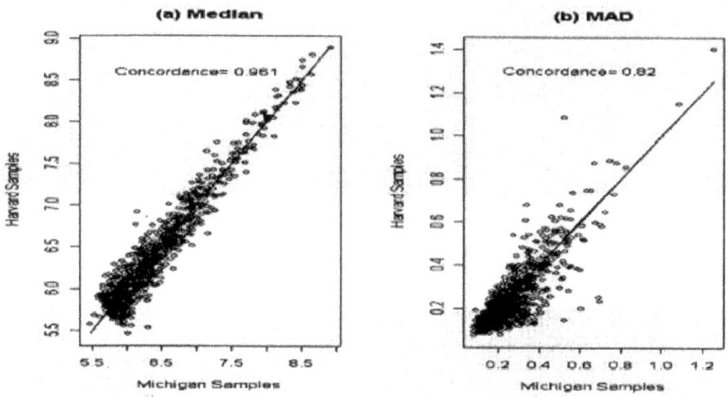

Figure 3. Median (a) and median absolute deviation (b) expression levels for each partial probeset based on the Harvard samples run on the U95Av2 chips vs. the Michigan samples run on the HuGeneFL chip. The high concordance in these measures suggests we obtain reasonably comparable expression levels by using the matched probes.

Recall that our method used only the matching probes, while completely ignoring expression level information for the non-matching probes. This means that our probesets are generally smaller than the Affymetrix-defined probesets. The median size of our "partial probesets" was seven, while the Affymetrix-defined probesets for the HuGeneFL and U95Av2 chips have 20 and 16 probes, respectively. Since additional probes can increase the precision in measuring the expression level of the corresponding gene, one might expect a loss of precision when using the partial probesets to quantify expression levels. To investigate this possibility, we quantified the expression levels for the full probesets of the Harvard samples using the PDNN model. The full probesets consisted of all probes on the array mapping to the Unigene cluster, i.e., not just the matching ones. We plotted the standard deviation for each gene using the full probeset versus the standard deviation for the partial probeset, given in Figure 4. If the partial probeset quantifications were considerably less precise, we would expect measurement error to cause the standard deviation to be larger for the partial probesets. There was no evidence of significant precision loss in this plot, as there is strong agreement between the standard deviations for each gene using the two methods (concordance=0.942). This may seem surprising at first, but upon further thought is reasonable, since we expect that the probes

Affymetrix chooses to retain in formulating new chips may be in some sense the "best" ones.

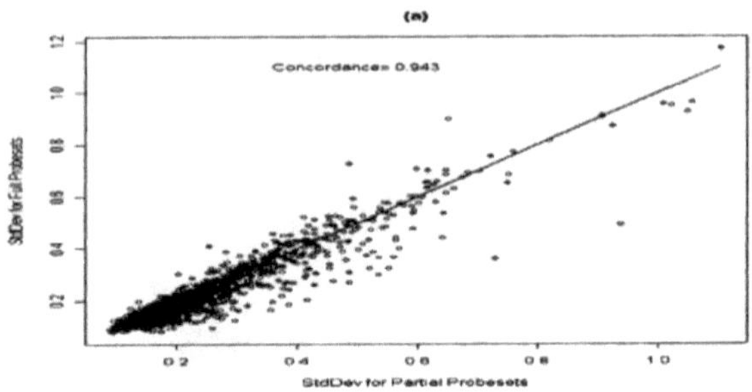

Figure 4. Standard deviation across Harvard samples for each gene based on full and partial probesets. A "full probeset" contains all probes on the U95Av2 chip mapping to a unique Unigene ID, while the corresponding "partial probeset" contains only the subset of probes contained on both the U95Av2 and HuGeneFL chips.

We computed Spearman correlations between the partial and full probeset quantifications for each probeset to confirm that our method preserved the relative ordering of the samples, i.e., the ranks. For example, we expect that a sample with the largest expression level for a given gene using the full set of probes will also demonstrate the largest expression level for that gene when using only the matched probes. The median Spearman correlation across all probesets was 0.95, suggesting that our method did a good job of preserving the relative ordering of the samples. Interestingly, but not surprisingly, most of the lower Spearman correlations occur for probesets with less heterogeneous expression levels across samples and/or probesets containing smaller numbers of probes. Thus, it appears that our partial probeset method worked quite well. We expect it to perform even better if it is used to combine information across U95 and U133 chips, since these chips share more probes in common than the HuGeneFL and U95 chips.

4. RESULTS

Figure 5(a) contains the histogram of permutation test p-values assessing the prognostic significance of each gene. The overabundance of probesets with very small p-values indicates the presence of some genes providing information on patient prognosis beyond what is offered by the modeled clinical factors. Table 1 contains a set of 26 genes that are flagged by the

BUM method using FDR<0.20, which are those genes with permutation p-values less than 0.0025. Our analogous BUM analyses found that 16 of these genes are also flagged based on the LRT, and 18 using the bootstrap. We also identified a set of genes that appear to be differentially expressed by clinical stage (early vs. late). Figure 5(b) contains the histogram of stage p-values from the Wilcoxon test, with the extreme right skewness indicating a very large number of significant genes. Using the BUM method with FDR<0.20, more than 1/3 of the genes (346/1036) were flagged as differentially expressed by stage. This was in contrast to the very small number (26) of genes flagged as prognostic with the same settings. This is not surprising, since one might expect that it is easier to identify genes related to an easily identifiable biological factor like stage than to predict how long the patient will live. There were 71 genes flagged using FDR<0.05, which corresponded to a p-value cutoff of 0.0064. Only one of the 26 genes we flag as prognostic is in the set of 71 genes flagged as related to stage using FDR<0.05 (STK25).

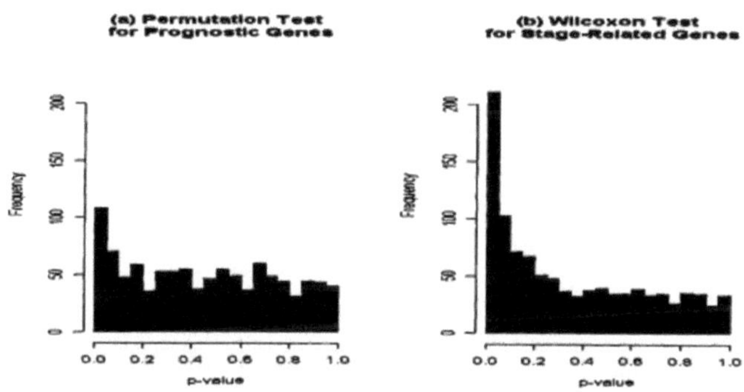

Figure 5. (**a**) Histogram of p-values from permutation test on gene coefficient in Cox model containing clinical covariates and each one of the 1036 candidate genes. The corresponding histogram for the LRT is nearly identical (**b**) Histogram of p-values from Wilcoxon test comparing median expression levels for early and late stage cancers.

5. INTERPRETATION OF RESULTS

We were able to link 10 of our 26 prognostic genes to lung cancer based on the existing literature. Four others could be linked to cancer in general or other lung disease in the literature. These genes are in boldface in Table 1.

Table 1. Set of genes flagged as prognostic by applying the BUM on the permutation p-values with FDR<0.20. Also included are the LRT and bootstrap p-values and estimates of the Cox model coefficient. A "*" indicates the p-value was below the BUM significance threshold. The identity of the genes is also given, with boldface type indicating we were able to find existing literature linking that gene with lung cancer, cancer in general, or other lung disease. A negative coefficient indicates that larger expression levels of that gene corresponded to a better survival outcome.

Gene Identity	Coef	Prognostic P-values		
		Permut.	LRT	Bootstrap
FCGRT; Fc fragment of IgG receptor	-2.07	<0.00001*	0.00014*	0.0006*
ENO2; Enolase 2	1.46	0.00001*	0.00002*	<0.0001*
NFRKB;Nuclear factor for kappaB binding	-2.81	0.00001*	0.00435	0.0040*
RRM1; Ribonucleotide reductase M1 polypeptide	1.81	0.00002*	0.00008*	<0.0001*
TBCE; Tubulin-specific chaperone e	-2.35	0.00004*	0.00069*	0.0006*
Similar to phosphoglycerate mutase 1	1.92	0.00008*	0.00020*	0.0004*
ATIC; IMP cyclohydrolase	1.81	0.00009*	0.00153*	0.0004*
CHKL; Choline kinase-like	-1.43	0.00010*	0.02305	0.0260
DDX3; DEAD/H box polypeptide 3	-2.37	0.00017*	0.00012*	0.0002*
OST; oligosaccharyltransferase	-1.64	0.00020*	0.00010*	0.0010*
CPE; Carboxypeptidase E	0.72	0.00031*	0.00053*	0.0010*
ADRBK1; Adrenergic, beta, receptor kinase 1	-2.20	0.00044*	0.00678	0.0030*
BCL9; B-cell CLL/lymphoma 9	-1.64	0.00067*	0.03602	0.0460
BZW1; Basic leucine zipper and W2 domains 1	1.33	0.00068*	0.00279*	0.0006*
TPS1; Tryptase, alpha	-0.64	0.00106*	0.00217*	<0.0001*
CLU; Clusterin	-0.52	0.00109*	0.00239*	0.0024*
OGDH; Oxoglutarate dehydrogenase	-2.19	0.00118*	0.00405	0.0020*
STK25; Serine/threonine kinase 25	2.29	0.00122*	0.00152*	0.0080
KCC2; potassium-chloride transporter 2	-1.70	0.00143*	0.00988	0.0220
SEPW1;Selenoprotein W, 1	-1.29	0.00145*	0.01026	0.0160
FSCN1; Fascin homolog 1, actin-bundling protein	0.66	0.00150*	0.00241*	0.0103
MRPL19; Mitochondrial ribosomal protn L19	1.12	0.00211*	0.03213	0.0340
ALDH9; Aldehyde dehydrogenase 9 family	-1.18	0.00223*	0.00378*	0.0020*
PFN2; Profilin 2	0.63	0.00248*	0.00351*	0.0020*
BTG2; BTG family, member 2	-0.75	0.00232*	0.00580	0.0140

Gene Identity	Coef	Prognostic P-values		
		Permut.	LRT	Bootstrap
FCGRT; Fc fragment of IgG receptor	-2.07	<0.00001*	0.00014*	0.0006*

The top gene in our list, FCGRT, is induced by Interferon γ in the treatment of SCLC [Pujol, et al., 1993]. The negative sign on our coefficient indicates that it is a positive prognostic factor, i.e., patients with high levels of this gene tend to have better prognoses. According to our model, every doubling of expression level of this gene corresponded to an 8-fold reduction in risk for death (hazard). RRM1 has been shown to be overexpressed in NSCLC, and one study found that patients with NSCLC who are treated with gemcitabine/cisplatin with low RRM1 mRNA levels show significantly longer survival times [Rosell, et al., 2003]. The positive sign on the regression coefficient indicates that our analysis also considered this gene to be a negative prognostic factor, meaning that higher expression levels corresponded to a poorer prognosis. Every doubling of the expression level corresponded with a 6-fold increase in the hazard.

Overexpression of selenoprotein W, 1 (SEPW1) has been shown to markedly reduce the sensitivity to H_2O_2 cytotoxicity in NSCLC cell lines [Jeong, et al., 2002]. This gene appeared as a positive prognostic factor in our analysis. FSCN1 has been demonstrated to be a prognostic marker of invasiveness in Stage I NSCLC [Pelosi, et al., 2003], and appeared as a negative prognostic factor in our analysis.

Some genes are lung cancer markers, either for NSCLC [CHKL, Ramirez de Molina, et al., 2002; ENO2, Ferrigno, Buccheri, and Giordano, 2003] or SCLC [CLU, Koyama, et al., 1998; CPE, North and Du, 1998]. ADBRK is co-expressed with Cox-2 in lung adenocarcinoma [Schuller, et al., 2001].

Some genes have been linked to other cancers. While it is possible that the connections between genes and lung cancer are circumstantial, we mention them here because some may be interesting and may turn out to be relevant to lung cancer. BCL9 is over-expressed in some cases of ALL [Katoh and Katoh, 2003], and NFRKB is amplified in AML. BTG2 has been demonstrated to inhibit cell proliferation in primary mouse embryo fibroblasts lacking functional p53, and is a positive prognostic gene in our analysis [Kuo, et al., 2003]. ATIC is a fusion partner of ALK that defines a subtype of anaplastic large cell lymphoma (ALCL) [Cheuk and Chan, 2001], and ALK itself has been linked with lung cancer. TPS1 is a unique protease, released from mast cell secretory granules into the respiratory tract of patients with inflammatory disease of the airways [Cairns and Walls, 1996].

None of the genes we identified appeared in the list of top 100 genes from the Michigan analysis [Beer, et al., 2002], and we have only found one (CPE) that was mentioned in the Harvard paper [Bhattacharjee, et al., 2001]. CPE was one of the genes defining a neuroendocrine cluster that they identified and associated with poor prognosis.

We repeated our analysis separately for the Harvard and Michigan data sets, i.e., without pooling, and only eight and one of the 26 genes, respectively, were flagged as having p-values less than 0.0024, while 17 are not flagged, including the top gene in our list (FCGRT). It is clear that we obtained significant gains by pooling information across the two studies.

6. DISCUSSION

It may seem curious that our list of prognostic genes had almost no overlap with the genes mentioned in other publications based on these data, but this is reasonable for several reasons. First, we addressed a different research question than the analyses done in those publications. We used multivariable Cox models to search for genes offering prognostic information *above and beyond* what has been provided by known clinical predictors. In the study of Beer, et al. [2002], researchers looked for prognostic genes, but they fit single-factor Cox models containing the gene expressions, but not clinical predictors. Thus, they were effectively searching for genes that provided information on survival, irrespective of whether the prognostic value of the gene was due to a possible association with known clinical factors like disease stage. Bhattacharjee, et al. [2001] approached the survival question indirectly by performing unsupervised clustering on the samples, testing which clusters had survival differences, then identifying the genes that were driving the clustering. Second, and perhaps more importantly, we gained increased power to detect prognostic genes as a result of pooling the data from the two studies.

There are clear benefits to be reaped by pooling information across microarray studies. Most microarray studies have small to moderate sample sizes, which means a relatively low statistical power that translates into a limited ability to detect significant relationships between gene expression levels and outcomes of interest. By pooling information across data sets, we can obtain additional sensitivity and specificity in identifying important genes. This may allow us to identify gene-outcome relationships that are undetectable in any one study alone. Of the 26 prognostic genes found in our analysis, 17 of them would not have been flagged by analogous methods in either the Harvard or the Michigan data without pooling. Given that many researchers make their data publicly available after publication, this suggests exciting possibilities for pooled analyses of existing data that could reveal important new insights into cancer biology.

Note that combining data across studies as we have done is fundamentally different from the pooling of results across studies that is typical in many meta-analyses. Pooling the data results in an increase in statistical power to detect differences, while simply pooling the results does not. However, one must be careful in combining data across studies. First, one must account for any study-to-study heterogeneity that may be caused by differences in the studies' patient populations or conditions. In this work, we dealt with this by incorporating a fixed effect for study in our survival models. If there are more than two studies available to pool, we recommend using either Bayesian hierarchical models (see Stangl [1996]) or frailty models (see Therneau and Grambsch [2000]), which both treat the study effect as random instead of fixed. These methods may not be as effective when pooling just two studies because they involve estimation of a variance component from a sample of size two.

Second, one must normalize the measurements to make them comparable across studies. In our case, this involves finding a way to effectively combine information across different microarray platforms. Here we have presented a new method that is applicable to oligonucleotide arrays in which we identify probes present on both platforms then combine them into new probesets based on Unigene clusters. Our investigations suggested that this method is reliable and precise, and yielded comparable gene expression quantifications across two different versions of Affymetrix chips, the HuGeneFL and HG-U95Av2, used in the Michigan and the Harvard studies. We expect that this method may perform even better in combining information across U95 and U133 chips, since these chips have more probes in common. We feel that this approach is stronger than simply trying to normalize the expression levels across chips using quantile normalization, for example, since it is actually extracting measurements from the arrays that have scientific reasons to be comparable, and not just trying to make an arbitrary adjustment on non-comparable measurements.

Our specific biological goal in this analysis was to identify prognostic genes, meaning genes that offered information on patient survival beyond what is provided by known clinical predictors. We accomplished this by fitting multivariable Cox models that contained the clinical predictors along with the genes. It is important to adjust for these factors, since a gene that is simply a surrogate for a known clinical predictor is not as useful to us since we can gain the prognostic information directly from the clinical predictor without the additional time and expense required to collect microarray data.

While this type of multivariable analysis may result in fewer flagged prognostic genes, we feel that this list has the potential to be more interesting biologically because we know that the flagged genes explain variability in patient survival not already explained by the clinical predictors. Many genes in our short list seem biologically interesting and have been linked with lung cancer in the existing literature.

7. CONCLUSIONS

We have introduced a method based on partial probesets that appears to be effective for combining expression data from different oligonucleotide arrays. Using this method, we have pooled information across the Harvard and Michigan studies and identified a set of genes that appear to be prognostic for lung adenocarcinoma, providing information above and beyond known clinical predictors. Many of these genes would not have been found without pooling, and a large proportion of them appear to be biologically interesting and are worthy of future investigation.

8. ACKNOWLEDGEMENTS

We would like to thank David Stivers and Kevin Coombes for helpful discussions regarding this analysis. We would also like to thank Lianchun Xiao and Sang-Joon Lee for their contributions on this project. We also thank the reviewers for their comments and suggestions.

9. REFERENCES

Beer, D, Kardia SLR Huang CC, Giordano TJ, Levin AM, Misek DE, Lin L, Chen G, Gharib TG, Thomas DG, Lizyness ML, Kuick R, Hayasaka S, Taylor JMG, lannettoni MD, Orringer MB, and Hanash S. Gene-expression profiles predict survival of patients with lung adenocarcinoma. *Nature Medicine* 8 (8): 816-24, 2002.

Benjamini, Y, and Hochberg, Y. Controlling the false discovery rate: a practical and powerful approach to multiple testing". *Journal of the Royal Statistical Society, Series B* 57(1): 289-300, 1995.

Bhattacharjee, A, Richards, WG, Staunton J, Li C, Monti S, Vasa P, Ladd C, Beheshti J, Bueno R, Gillette M, Loda M, Weber G, Mark EJ, Lander ES, Wong W, Johnson BE, Golub TR, Sugarbaker DJ, and Meyerson M. Classification of human lung carcinomas by mRNA expression profiling reveals distinct adenocarcinoma subclasses. PNAS 98 (24), 13790-13795, 2001.

Cairns JA, Walls AF. Mast cell tryptase is a mitogen for epithelial cells. Stimulation of IL-8 production and intercellular adhesion molecule-1 expression. *Journal of Immunology* 156(1): 275-83, 1996.

Cheuk W, Chan JK.. Timely topic: anaplastic lymphoma kinase (ALK) spreads its influence. *Pathology* 33(1):7-12, Review, 2001.

Curran JE, Vaughan T, Lea RA, Weinstein SR, Morrison NA, Griffiths LR. Association of A vitamin D receptor polymorphism with sporadic breast cancer development. *International Journal of Cancer* 3(6):723-6, 1999.

Ferrigno D, Buccheri G, Giordano C. Neuron-specific enolase is an effective tumour marker in non-small cell lung cancer (NSCLC). *Lung Cancer* 41(3):311-20, 2003.

Jeong D, Kim TS, Chung YW, Lee BJ, Kim IY. Selenoprotein W is a glutathione-dependent antioxidant in vivo. *FEBS Letters* 517(1-3):225-8, 2002.

Katoh M, Katoh M. Identification and characterization of human BCL9L gene and mouse Bcl9l gene in silico. *International Journal of Molecular Medicine* 12(4):643-9, 2003.

Koyama Y, Yang HM, Wargalla U, Reisfeld RA, Harper JR. Biochemical characterization of a sulfated phosphoglycoprotein antigen expressed on human small cell lung carcinoma. *Journal of Biological Chemistry* 263(2):806-11, 1998.

Kuo ML, Duncavage EJ, Mathew R, den Besten W, Pei D, Naeve D, Yamamoto T, Cheng C, Sherr CJ, Roussel MF. Arf induces p53-dependent and -independent antiproliferative genes. *Cancer Research* 63(5):1046-53, 2003.

Schadt EE, Li C, Ellis B, Wong WH, Feature extraction and normalization algorithms for high-density oligonucleotide gene expression array data. *J Cell Biochem* Suppl. Suppl 37:120-5, 2001.

North WG and Du J. Key peptide processing enzymes are expressed by a variant form of small-cell carcinoma of the lung. *Peptides* 19(10): 1743-7, 1998.

Pelosi G, Pasini F, Sonzogni A, Maffini F, Maisonneuve P, lannucci A, Terzi A, De Manzoni G, Bresaola E, Viale G. Prognostic implications of neuroendocrine differentiation and hormone production in patients with Stage I nonsmall cell lung carcinoma. *Cancer* 97(10):2487-97, 2003.

Pounds, S and Morris, S. Estimating the occurrence of false positives and false negatives in microarray studies by approximating and partitioning the empirical distribution of p-values". *Bioinformatics,* 19, 1236—1242, 2003.

Pujol JL, Gibney DJ, Su JQ, Maksymiuk AW, Jett JR. Immune response induced in small-cell lung cancer by maintenance therapy with interferon gamma. *Journal of the National Cancer Institute* 85(22): 1844-50, 1993.

Ramirez de Molina A, Rodriguez-Gonzalez A, Gutierrez R, Martinez-Pineiro L, Sanchez J, Bonilla F, Rosell R, Lacal J. Overexpression of choline kinase is a frequent feature in human tumor-derived cell lines and in lung, prostate, and colorectal human cancers. *Biochemical and Biophysical Research Communications* 296(3):580-3, 2002.

Rosell R, Crino L, Danenberg K, Scagliotti G, Bepler G, Taron M, Alberola V, Provencio M, Camps C, De Marinis F, Sanchez JJ, Penas R. Targeted therapy in combination with gemcitabine in non-small cell lung cancer. *Seminars in Oncology.* 30(4 Suppl 10): 19-25. Review, 2003.

Schuller HM, Plummer HK 3rd, Bochsler PN, Dudric P, Bell JL, Harris RE. Co-expression of beta-adrenergic receptors and cyclooxygenase-2 in pulmonary adenocarcinoma. *International Journal of Oncology* 19(3):445-9, 2001.

Stangl, DK. Hierarchical Analysis of Continuous-Time Survival Models. *Bayesian Biostatistics,* DA Berry and DK Stangl, eds., Marcel Dekker, New York: 429-450, 1996.

Therneau, TM and Grambsch, PM. *Modeling Survival Data: Extending the Cox Model.* Springer, New York, 2000.

Zhang, L, Coombes, K, and Xia, L. Quantifications of Cross Hybridization on Oligonucleotide Microarrays. *Methods of Microarray Data Analysis III,* Lin, S., and Johnson, K., eds., Kluwer, New York, 2003.

Zhang, L, Miles, MF, Aldape, KD. A model of molecular interactions on short Oligonucleotide microarrays. *Nature Biotechnology* 21(7): 818-21, 2003.

Chapter 5

APPLICATION OF SURVIVAL AND META-ANALYSIS TO GENE EXPRESSION DATA COMBINED FROM TWO STUDIES

Linda Warnock, Richard Stephens, JoAnn Coleman
GlaxoSmithKline

Abstract: The application of gene expression microarray technology has the potential to have a large impact in the area of oncology. There is a need to be able to identify genes associated with prolonged or reduced survival, to aid decisions regarding patient treatment and care. In addition these genes can be targeted in drug research to aid discovery and development of novel treatments. This paper uses two published Affymetrix datasets and combines the information from adenocarcinoma lung tumors to identify genes associated with survival. Kaplan-Meier survival analysis, Cox proportional hazards models and analysis of variance are used for the data analyses. The results are combined across the two datasets using Fisher's chi-squared meta-analysis based on p-value aggregation. The false discovery rate (FDR) adjustment is made to the final p-values.

Key words: Affymetrix, principal component analysis, Kaplan-Meier analysis, Cox proportional hazards model, meta-analysis, false discovery rate, lung adenocarcinoma

1. INTRODUCTION

Gene expression data and clinical information have been collected from two experiments designed to investigate the relationship between gene expression and survival in patients with lung cancer. Expression data has been collected from studies run in association with Harvard [Bhattacharjee et al., 2001] and Michigan [Beer et al., 2002] universities. The two studies used different Affymetrix chip types to produce the gene expression data and have collected a variety of clinical information. This analysis combines the information across these two datasets and focuses on lung adenocarcinoma tumors to identify genes which are associated with survival.

2. METHODS

2.1 Combining Information

The data were downloaded from the CAMDA website [www.camda.duke.edu/camda03/] in the form of .CEL files. The .CEL files contain the raw intensity data prior to normalization. The intensity data had been generated from a total of 299 tumor samples (Harvard: 203, Michigan: 96). The Harvard data collected samples from adenocarcinoma (139), squamous (21), small cell lung cancer (six), carcinoid (20) and normal (17) tumors; however only 124 of the adenocarcinoma samples had clinical information in addition to gene expression data. The Michigan data had 86 adenocarcinoma samples all with associated clinical information and gene expression data.

The combining of the data across the two studies was complicated by the fact that the expression data had been collected using different Affymetrix chip technologies. The Harvard study used the U95A chip with 16 probe pairs per probe set and 12,625 probe sets while the Michigan study used the older HuGene FL chip with 20 probe pairs per probe set and 7,129 probe sets. The probe sets on the two chips are designed differently and do not always target the same genes. Hence the two chips had a mixture of common genes and completely different genes represented. The common genes may also be represented by probe sets targeting different parts of the gene sequence. The Affymetrix website provides comparison spreadsheets which allows probe sets, targeting the same gene, to be matched from different chip types. The HuGeneFL_to_U95_comp.xls spreadsheet [www.affymetrix.com/support/technical/comparison_spreadsheets.affx] was used to select probe sets which had a sequence relationship between the two chips. This matching resulted in over 6,000 probe sets being defined as common between the two datasets. This method of matching is more precise than using gene names. The set of common probe sets (genes) was used in all subsequent analyses.

2.2 Pre-processing of the data

The Harvard and Michigan gene expression data were pre-processed and normalized independently of each other. The .CEL files containing the raw expression data were processed in MAS5 [www.affymetrix.com/products/software/specific/mas.affx] and DChip [Li and Wong, 200la] software. The quality control (QC) process involved identification of chip to chip variation using DChip and MAS5 algorithms. Any chips with 'probe set outlier %' greater than three (DChip) were discarded. B-actin and Gapdh were housekeeper genes represented on every chip. Any chip with the 3'/5' ratios greater than three for the housekeepers were also discarded (MAS5). Metrics related to the background intensity and overall chip intensity such as 'raw Q', 'scale factor', 'background intensity' were collected and used in a principal components analysis (PCA) in SIMCA-P+ version 10 [www.umetrics.com/software_simcapplus.asp] with the aim of identifying poor quality chips. Using these quality control criteria, 20 chips were removed from the Harvard dataset (10 of which were adenocarcinoma samples) and 16 chips from the Michigan dataset. The Harvard data contained information on the *in-vitro* transcription (IVT) batch which was used in the process of sample preparation. Through PCA analysis of the Harvard QC data it was found that one batch (28 chips) of IVT produced an overall lower average chip signal and lower background signal. This created some bias in the expression data; however the effect did not appear great enough to justify the removal of 28 chips. This finding showed the importance of identifying technological variation and ideally repeating the chip hybridizations.

The two datasets were normalized in DChip using the piece-wise linear normalization algorithm on the perfect match (PM) data only. All of the data exploration and analyses were performed on the PM data only. The data could not be combined at this stage due to the different chip types (HuGene FL: Michigan, U95A: Harvard).

2.3 Exploratory analysis of the clinical data

The effect of the variables sex, age and tumor stage on survival were investigated using a Kaplan-Meier plot and Cox proportional hazards regression model. The cancer staging handbook [Greene, 2002] was used to classify each tumor as stage I (72), II (22), III (eight) or IV (11) in the Harvard dataset (one sample did not have a stage classification), with stage IV being metastasis or development of a secondary tumor. The Michigan dataset only had patients with stage I (56) or stage III (14) tumors. The

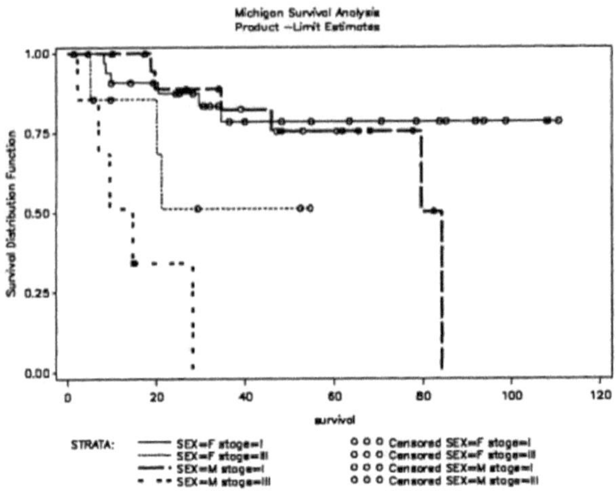

Figure 1. Kaplan-Meier plot of the Michigan clinical data showing the effect of tumor stage and sex on survival. Stage I tumors have a greater survival rate than stage III.

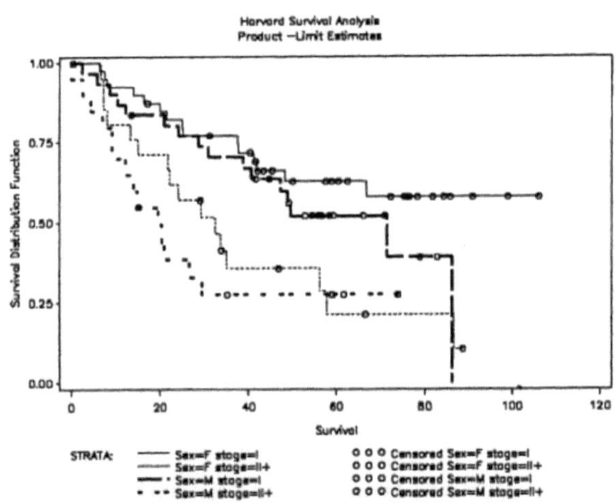

Figure 2. Kaplan-Meier plot of the Harvard clinical data showing the effect of tumor stage and sex on survival. As seen for the Michigan data stage I tumors have a greater survival rate than stage II+.

significance of the variables were determined using a forward selection Cox regression model in SAS version 8 [Allison, 1995]. The analysis was performed separately on each dataset, 114 Harvard adenocarcinoma samples and 70 Michigan adenocarcinoma samples. Each sample was represented by over 6,000 genes which were identified as common between the two datasets. Figures 1 and 2 showed that stage of tumor had a large effect on survival rates with stage I having a greater survival rate. The significance of the clinical variables were investigated with a Cox proportional hazards regression model. This confirmed that stage had a large, significant effect on survival (Harvard p=0.0003, Michigan p<0.0001). There was slight evidence of a difference between sex (Harvard p=0.4074, Michigan p=0.0362) and slight evidence of an effect of age (Harvard p=0.2398, Michigan p=0.0526).

2.4 Exploratory Analysis of the expression data

Principal component analysis (PCA) was used to explore the quality controlled, chip-normalized expression data to look for large sources of variation across all the chips (Harvard and Michigan). The scores plot (Figure 3) plotted the first principal component on the y-axis. This component accounted for 89% of the total variation and showed the separation of the expression intensity between Michigan and Harvard. There were several possible reasons for this difference which are outlined in the Section 4.

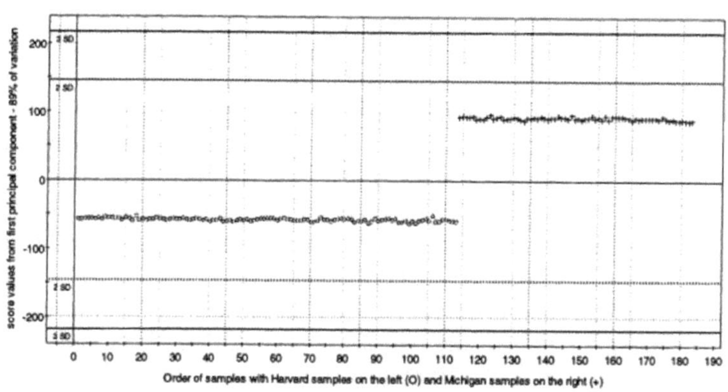

Figure 3. PCA scores plot showing the separation between Harvard and Michigan gene expression data (crosses = Michigan samples, circles = Harvard samples). The first component is shown on the y-axis and accounts for 89% of the variation. The x-axis is simply an ordering of samples.

The loadings (not shown) for the first component showed that the majority of genes had high, positive loadings indicating that the majority were more highly expressed for one dataset than the other, although there were a small number of genes where the converse was true.

Summary statistics showed the geometric mean expression for Harvard to be 2.4 (raw average intensity of 250) with a standard deviation of 0.50 and for Michigan to be 3.1 (raw average intensity of 1260) with a standard deviation of 0.38 demonstrating an overall increase in intensity for the Michigan expression data.

2.5 Identification of genes associated with survival

Cox proportional hazards regression model was used to identify genes associated with survival. The effect of sex, age and tumor stage were assessed in addition to log expression intensity. This analysis was performed for every gene and a ranking based on the statistical significance of each gene was used to identify the genes most associated with survival.

A forward selection process was used so that variables were entered in order of greatest association with survival. A p-value entry of 1 was used to ensure that every variable was entered into the model. Using this approach the significance of each variable was assessed after accounting for the previous variable entered. If two variables were correlated with each other, and also with survival, then only one of the variables was necessary for survival prognosis as the prognosis effect of the second variable will be negligible after taking into consideration the first variable. In this analysis tumor stage was usually the most highly correlated with survival and hence was entered first into the model. Only five genes from the Michigan data and one from the Harvard data showed the gene intensity variable as more important than the stage variable. Two of these genes showed a significant association with survival in both datasets (probe sets 34777_at and 40507_at). These genes can be found in Table 1.

An important aspect of the analysis was the combination of results across the two datasets. Although the two datasets could have been normalized to remove the large source of variation between them, it was decided to keep the two datasets separate and combine the results using the chi-squared meta-analysis method developed by Fisher [1932] and applied by Rhodes et al. [2002] to gene expression. This method allowed the two datasets to be analyzed separately and the final p-values to be aggregated into a new meta-analysis p-value using chi-squared distribution theory.

A false discovery rate adjustment [Benjamini and Hochberg, 1995] was used on the meta-analysis p-values across all the genes. The volcano plots [Wolfinger et al., 2001] in Figures 4 and 5 summarize the results for the two datasets prior to p-value aggregation. Minus log base 10 of the p-value was plotted against the parameter estimate of the expression intensity for every

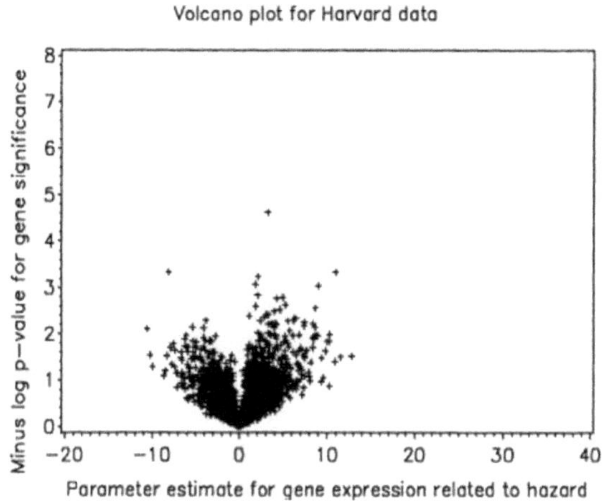

Figure 4. Volcano plot summarizing the Cox regression results from the Harvard dataset for all the genes. Minus log base 10 of the p-value was plotted against the parameter estimate of the expression intensity for every gene thus allowing the size of the effect to be assessed alongside the statistical significance. Values of 1.3, 2 and 3 on the y-axis correspond to p-valuesof 0.05, 0.01 and 0.001 respectively.

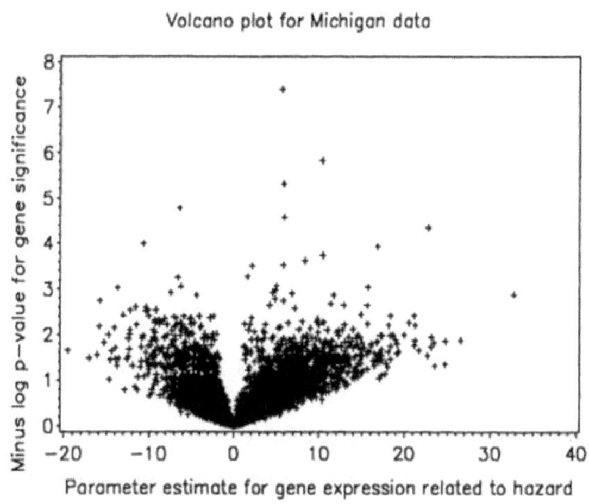

Figure 5. Volcano plot summarizing the Cox regression results from the Michigan dataset for all the genes. Minus log base 10 of the p-value was plotted against the parameter estimate of the expression intensity for every gene thus allowing the size of the effect to be assessed alongside the statistical significance. Values of 1.3, 2 and 3 on the y-axis correspond to p-values of 0.05, 0.01 and 0.001 respectively.

gene. The parameter estimate of the expression intensity was taken from the Cox regression analysis and can be thought of as an indication of the biological significance of the gene. The larger the parameter estimate (positive or negative) the greater the association with survival (decrease or increase respectively). The exponential of the parameter estimate can be interpreted as the increase in hazard (risk of death) for every unit increase in log gene expression or in other words the increase in hazard for every 10-fold increase in gene expression.

The volcano plots showed that the Michigan results (Figure 5) had a larger range of effect with the parameter estimates ranging from -20 to 35 whereas the range of effect in the Harvard data was from -10 to 15. The Michigan results also showed a greater number of significant genes. Any gene which had a positive association in one dataset but negative in the other were discarded. Two hundred and forty-one genes had a meta-analysis $p \leq 0.05$, 67 with meta-analysis $p \leq 0.01$, nine with meta-analysis $p \leq 0.001$ (shown in Table 1) and two with a FDR adjusted meta-analysis $p \leq 0.05$ (Table 1).

2.6 Identification of genes associated with tumor stage

An alternative analysis approach used the tumor stage as a surrogate marker for survival as patients with a stage I tumor were more likely to survive than patients with more advanced tumors. This approach used analysis of covariance (ANCOVA) with log gene expression as a response, tumor stage (classified as stage I or stage II+) and sex as explanatory variables and age as a covariate. This approach addressed the same problem of identifying genes associated with survival but from a different angle. The previous analysis identified genes which were associated with survival after accounting for the effect of tumor stage whereas this analysis looked directly for genes associated with tumor stage (acting as a surrogate for survival). As a result the genes identified were likely to be different from the previous analysis, however there were some overlapping genes mentioned below. Fisher's meta-analysis was also used with this approach to combine the p-values from the two datasets and the FDR p-value adjustment was made.

It is interesting to note that there were very few genes with a two-fold change or greater. A filter was placed on the genes so that only genes with a 1.5 fold change in at least one dataset were considered meaningful. This resulted in 43 genes with meta-analysis $p \leq 0.05$, 36 with meta-analysis $p \leq 0.01$, 27 with meta-analysis $p \leq 0.001$, 22 with FDR adjusted meta-analysis $p \leq 0.01$ (shown in Table 2) and 32 with FDR adjusted meta-analysis $p \leq 0.05$. There were four genes with meta-analysis $p \leq 0.05$ which also had meta-analysis $p \leq 0.05$ from the Cox analysis (201_s_at, 36780_at, 37006_at,

37394_at). Genes 201_s_at and 37394_at appear in Table 2 but were not significant enough to appear in Table 1. Genes 36780_at and 37006_at were not significant enough to appear in either table.

The genes which had a significant effect on survival were looked at in more detail by using gene ontology (GO) [www.geneontology.org], discussed in Section 3.

3. RESULTS

Table 1 showed a subset of the results of the Cox regression analysis by providing a list of the nine most significant genes in descending order of meta-analysis p-value. The genes were potential prognostic markers of survival with meta-analysis $p \leq 0.001$ (Cox analysis), however only two of the genes had FDR adjusted meta-analysis $p \leq 0.05$. The parameter estimate for gene intensity gave an indication of the size of association between gene intensity and survival. A value of 1 implied that the risk of death increased by approximately 2.7 times (exponential of 1) for a 10-fold increase in gene expression. Table 2 showed a subset of the results of the ANCOVA analysis by providing a list of the 22 most significant genes in descending order of meta-analysis p-value. Only genes which had more than a 1.5 fold-change in at least one of the datasets were listed. The genes were potential markers of tumor stage and hence of survival with FDR adjusted meta-analysis $p \leq 0.01$. There were many significant genes however few had a meaningful fold-change between stage I and stage II+. The first gene in Table 2, CD37 antigen, had a negligible fold-change in the Harvard dataset and a -1.57 fold-change in the Michigan dataset. The negative sign indicates down-regulation which means that the gene expression intensity has decreased from stage II+ to stage I.

Table 1. List of genes, with meta-analysis p-value ≤ 0.001 identified by Cox regression analysis as prognostic for survival. The FDR adjusted meta-analysis p-value, the hazard parameter estimates and the standard error (SE) of the estimate for each gene are given. Genes highlighted in italics have appeared in cancer literature.

probe_set	meta-analysis p-value	FDR adjusted p-value	Harvard intensity parameter estimate (SE)	Michigan intensity parameter estimate (SE)	Gene name
34777_at	2.48E-07	0.0015	1.00 (0.9)	5.67 (2.2)	*adrenomedullin*
40507_at	2.59E-06	0.0077	8.07 (3.6)	10.34 (2.8)	*solute carrier family 2 (facilitated glucose transporter), member 1*
1649_at	6.33E-05	0.0656	5.68 (3.3)	22.78 (6.2)	chromosome 20 open reading frame 16
32300_s_at	0.0002	0.1582	5.23 (1.7)	13.99 (5.6)	*tyrosine hydroxylase*
38544_at	0.0003	0.2286	2.10 (0.8)	4.99 (2.4)	*inhibin, alpha*
1269_at	0.0005	0.3053	-2.71 (1.1)	-7.04 (2.2)	*phosphoinositide-3-kinase, regulatory subunit, polypeptide 1*
35693_at	0.0006	0.3195	2.48 (3.4)	16.87 (4.3)	hippocalcin-like 1
36133_at	0.0007	0.3428	0.46 (0.4)	8.33 (2.6)	*desmoplakin (DPI, DPII)*
32593_at	0.0009	0.3680	-0.17 (0.8)	-10.60 (2.6)	KIAA0084 protein

Table 2. (continued on next page). List of genes, with FDR adjusted meta-analysis p ≤ 0.01 identified by ANCOVA as prognostic for tumor stage. The fold-change estimates indicate the size of the difference between stage I and stage II+ tumors with associated 95% confidence intervals.

probe_set	meta-analysis p-value	FDR adjusted p-value	Harvard fold change between stages (95% confidence interval)	Michigan fold change between stages (95% confidence interval)	Gene name
31870_at	1.25E-09	7.48E-06	-1.06 (-1.12, -1.01)	-1.57 (-1.79, -1.79)	CD37 antigen
1288_s_at	3.54E-07	0.0003	-1.06 (-1.12, -1.01)	-1.51 (-1.75, -1.30)	J04617 /FEATURE=cds /DEFINITION=HUMEF1A Human elongation factor EF-1-alpha gene, complete cds
31962_at	1.55E-06	0.0006	-1.03 (-1.1, -0.97)	-1.54 (-1.78, -1.32)	ribosomal protein L37a
32466_at	5.34E-06	0.0011	-1.01 (-1.05, -0.97)	-1.61 (-1.91, -1.36)	ribosomal protein L41
36792_at	6.89E-06	0.0013	1.65 (1.37, 2.00)	1.07 (0.79, 1.43)	tropomyosin 1 (alpha)

probe_set	meta-analysis p-value	FDR adjusted p-value	Harvard fold change between stages (95% confidence interval)	Michigan fold change between stages (95% confidence interval)	Gene name
37892_at	9.69E-06	0.0016	1.79 (1.34, 2.37)	1.64 (1.15, 2.32)	collagen, type XI, alpha 1
1385_at	2.91E-05	0.0027	1.70 (1.31, 2.19)	1.35 (1.04, 1.76)	transforming growth factor, beta-induced, 68kDa
38111_at	3.30E-05	0.0028	1.70 (1.24, 2.32)	1.71 (1.22, 2.40)	chondroitin sulfate proteoglycan 2
1237_at	3.42E-05	0.0028	1.5 (1.25, 1.81)	1.27 (0.95, 1.71)	immediate early response 3
1179_at	4.25E-05	0.0031	1.01 (0.91, 1.12)	1.56 (1.31, 1.86)	Heat Shock Protein, 70 Kda
31775_at	5.79E-05	0.0039	-1.5 (-1.86, -1.21)	-1.53 (-2.13, -1.10)	Cluster Incl. X65018:H.sapiens mRNA for lung surfactant protein D
32305_at	6.84E-05	0.0041	1.94 (1.35, 2.77)	1.63 (1.11, 2.39)	collagen, type I, alpha 2
34760_at	7.60E-05	0.0043	-1.52 (-1.87, -1.23)	-1.27 (-1.61, -1.00)	KIAA0022 gene product
658_at	7.81E-05	0.0043	1.92 (1.41, 2.63)	1.34 (0.96, 1.88)	thrombospondin 2
37004_at	0.0001	0.0051	-2.2 (-3.23, -1.50)	-1.57 (-2.70, -0.92)	surfactant, pulmonary-associated protein B
201_s_at	0.0001	0.0051	-1.02 (-1.14, -0.92)	-1.59 (-1.93, -1.31)	beta-2-microglobulin
39337_at	0.0001	0.0056	1.15 (0.98, 1.35)	1.57 (1.26, 1.96)	H2A histone family, member Z
35730_at	0.0001	0.0060	-1.25 (-1.46, -1.07)	-1.92 (-2.89, -1.27)	alcohol dehydrogenase IB (class I), beta polypeptide
33754_at	0.0001	0.0060	-2.18 (-3.08, -1.55)	-1.09 (-1.61, -0.74)	thyroid transcription factor 1
39945_at	0.0003	0.0093	1.58 (1.23, 2.03)	1.28 (0.99, 1.66)	fibroblast activation protein, alpha
37394_at	0.0003	0.0094	-1.43 (-1.88, -1.09)	-1.74 (-2.46, -1.22)	complement component 7
38744_at	0.0003	0.0094	1.1 (0.98, 1.24)	1.55 (1.24, 1.94)	Deleted in split-hand/split-foot 1 region

These analyses have identified a number of genes as being associated with survival, either through Cox regression analysis or by using ANCOVA. A literature search was performed on the most significant genes obtained from the Cox regression analysis. Of the top nine genes, six were found in literature searches to be related to cancer and these genes have been highlighted in italics in Table 1. The Cox analysis (Table 1) showed that adrenomedullin had a negative association with survival (positive parameter estimates) with an estimate of 1 in the Harvard dataset and 5.67 in the Michigan dataset. This implied that the risk of death increased by approximately 2.7 and 290 times respectively for every 10-fold increase in intensity. It is questionable whether an increase in risk of 2.7 times for a 10-fold increase in expression was big enough to be biologically meaningful. This gene was found in a literature search to be an 'important tumor survival factor in human carcinogenesis' [Cuttitta et al., 2002]. The only other gene which had a significant FDR adjusted meta-analysis p-value was solute carrier, glucose transporter. This gene was one of the few to show consistent results between the two datasets with parameter estimates of 8.07 and 10.34 (Table 1). The analysis of covariance showed many genes to be statistically significant but fewer to be biologically significant. The first gene in Table 2 to show a change in gene expression between stage I and stage II+ is collagen, type XI with fold changes of 1.79 (Harvard) and 1.64 (Michigan).

The tables present the combined meta-analysis p-value alongside the two individual study effects. Showing the effects from the two studies provided the opportunity of assessing the biological agreement between Harvard and Michigan results. If the data had been aggregated prior to analyses this assessment could not have taken place. Over all the genes, the analysis showed very little agreement of results. The Michigan data tended to give more favourable results from the Cox analysis with larger parameter estimates and smaller p-values (as seen by comparing Figures 4 and 5).

Gene ontology (GO) is a way of assessing a gene's function in three areas: the biological process, the cellular component and the molecular function. The 241 genes identified from Cox regression with meta-analysis $p \leq 0.05$ were investigated for GO groupings. Thirty-one genes were involved in the biological process of signal transduction, 14 in oncogenesis, 10 in immune response, 10 in inflammatory response, 10 in cell proliferation, nine in cell motility and eight in cell-cell signalling, which included adrenomedullin. The molecular function assessment gave 17 genes involved in transcription factor, nine in DNA binding, eight in cell adhesion and eight in protein binding. The cellular component assessment gave 32 genes active in the integral plasma membrane protein, 18 in the plasma membrane, 14 in the nucleus, eight in the cytoplasm and eight in the extracellular space.

4. DISCUSSION

The task of combining data across chip types and different data sources was challenging. The PCA plot (Figure 1) demonstrated heterogeneity between the data sets. Some possible explanations for this clear separation are different scanning intensities used for the different chip types, different methods used for processing the data within the two sites or differences between the probe sets across the Harvard and Michigan datasets used to target the same sequences. This paper showed how a meta-analysis technique could be used to take this issue into account instead of the more normal approach of normalizing across the datasets prior to analyses. It allowed the results of the two datasets to be assessed both independently and together, and can be thought of as using the results from one dataset to validate the results from the other.

Two approaches were used to analyze the data: Cox regression and analysis of covariance. The Cox method was essential when using censored survival times as there were in these studies. As expected, the different approaches identified different gene sets but both methods were useful for identifying genes to aid decisions regarding patient care and to aid discovery and development of novel treatments. Overall there was little agreement between the results from the two datasets and so it was difficult to put too much faith on the results without further validation or follow-up work on the genes identified. A possible reason for the lack of agreement is the variability introduced from the tumor samples. Very little was known about the collection of the samples or about the patients. Sources of variability could include: patient treatment, date of the collection of clinical information, date of collection of gene expression data, date of prognosis, date of dissection, quality of hospital resources such as equipment and training of staff, race of patient, occupation of patient. These are just a subset of variables which could add variability thus overshadowing changes in survival due to gene expression. It would not be possible to account for all these variables in an analysis but these variables could be taken into consideration when recruiting patients to take part in a study.

ACKNOWLEDGEMENTS

We would like to thank Robert Gagnon for his thorough review of the paper, his helpful comments and support throughout this project, David Willé for his discussions on meta-analysis, Lini Pandite for her scientific input on issues concerning lung adenocarcinomas, Priti Hegde and Chang Liu for their help with gene ontology.

REFERENCES

Allison P (1995) Survival Analysis Using the SAS System: A Practical Guide. Cary NC: SAS Institute Inc.

Beer D, Kardia S, Huang C, Giordano T, Levin A, Misek D, Lin L, Chen G, Gharib T, Thomas D, Lizyness M, Kuick R, Hayasaka S, Taylor J, Iannettoni M, Orringer M and Hanash S (2002) Gene-expression profiles predict survival of patients with lung adenocarcinoma. Nature Medicine 8: 816-824

Benjamini, Y and Hochberg, Y (1995) Controlling the false discovery rate: a practical and powerful approach to multiple testing. Journal of the Royal Statistical Society Series B, 57(1): 289-300.

Bhattacharjee A, William G, Richards W, Staunton J, Li C, Monti S, Vasa P, Ladd C, Beheshti J, Bueno R, Gillette M, Loda M, Weber G, Mark E, Lander E, Wong W, Johnson B, Golub T, Sugarbaker D, and Meyerson M (2001) Classification of human lung carcinomas by mRNA expression profiling reveals distinct adenocarcinoma subclasses. PNAS 98: 13790-13795.

Cuttitta F, Pío R, Garayoa M, Zudaire E, Julián M, Elsasser T, Montuenga L and Martínez A (2002) Adrenomedullin functions as an important tumor survival factor in human carcinogenesis. Microscopy Research and Techique 57(2): 110-9.

Fisher R (1932) Statistical Methods for Research Workers, Forth Edition. London: Oliver and Boyd.

Greene F (2002) AJCC Cancer Staging Handbook, Sixth Edition, Springer Verlag: 191-203

Li C and Wong W (2001a) Model-based analysis of oligonucleotide arrays: Expression index computation and outlier detection. Proc. Natl. Acad. Sci. 98: 31-36.

Rhodes D, Barrette T, Rubin M, Ghosh D and Chinnaiyan A (2002) Meta-Analysis of Microarrays: Interstudy Validation of Gene Expression Profiles Reveals Pathway Dysregulation in Prostrate Cancer. Cancer Research 62: 4427-4433.

Wolfinger R, Gibson G, Wolfinger E, Bennett L, Hamadeh H, Bushel P, Afshari C and Paules R (2001) Assessing gene significance from cDNA microarray expression data via mixed models. Journal of Computational Biology, 8(6): 625-637.

Chapter 6

MAKING SENSE OF HUMAN LUNG CARCINOMAS GENE EXPRESSION DATA: INTEGRATION AND ANALYSIS OF TWO AFFYMETRIX PLATFORM EXPERIMENTS

Xiwu Lin, Daniel Park, Sergio Eslava, Kwan R. Lee, Raymond L.H. Lam, and Lei A. Zhu

Biomedical Data Sciences. GlaxoSmithKiline, 1250 S. Collegeville Rd.,Collegeville,PA 19426

Abstract: High throughput technologies such as microarray, mass spectrometry and nuclear magnetic resonance, have generated large volumes of valuable data for biology research. Researchers often face the challenges of integrating data from different sources and of identifying potential biomarkers that are highly associated with disease, drug safety, and efficacy. We present several solutions to these challenges through two Affymetrix microarray studies aimed at providing new insights into lung cancer biology. The Harvard dataset and the Michigan dataset were integrated to identify genes that were predictive of cancer survival. Quantile normalization of expression measures was applied to make the two datasets comparable. Genes highly associated with survival were identified and survival tree analysis on the combined data was performed to predict mortality. The candidate genes could be useful for lung cancer disease prediction and cancer therapy. The methodologies for integration and analysis of multiple gene expression data have been shown to perform well and could be generalized to broader applications.

Key words: Gene expression, integration, Affymetrix MAS, principal component analysis, partial least squares, survival tree

1. INTRODUCTION

The data sets for the 2003 CAMDA focused on lung cancers. Four microarray platform data sets were released for integration and combined analysis. In this paper we present several solutions to integrate the Harvard [Bhattacharjee et al., 2001] and the Michigan [Beer et al., 2002] data sets and to identify potential biomarkers that are highly associated with lung cancer.

The two platform data sets were independently acquired from two different studies that used different sets of samples and two different Affymetrix gene chips. There are several challenges in integrating across the two platforms. Firstly, appropriate data processing is required to make the raw data (ie the probe level data in the form of CEL files) ready for analysis. There are issues with the existing processed data such as negative values and large variability among the low expressed genes. It is possible that the data were generated using an earlier version of Affymetrix MAS software (version 4.0). Re-processing the probe level data with a newer version MAS 5.0 [Affymetrix, 2001] will overcome these issues. Secondly, the two different chips lead to two different sets of genes (clones). Merging the two platforms by gene names alone could result in very few common genes. An alternative approach is to merge the two platforms by probe set ID, using the Affymetrix array comparison spreadsheet (www.Affymetrics.com). Thirdly, data for the same genes are not comparable across studies. We modify the quantile normalization to make samples comparable across different platforms for each gene. The modification aims to remove differences due to different platforms.

In Section 2 we compare the original processed data and the MAS5.0 data, evaluate two ways of merging the two platforms, and describe the modified quantile normalization method. In Section 3 we examine the integrated data using PCA (principal component analysis) and PLS-DA (partial least squares discriminant analysis). In Section 4 we evaluate the validity of the integration and the prediction performance of several data mining methods. This is done by treating the Harvard data as a training set and the Michigan data as a test set in the discrimination of normal lung samples from the adenocarcinomas. In Section 5, we identify genes highly associated with survival and perform survival tree analysis on the integrated data to predict mortality. Section 6 provides some conclusions and a discussion of further work.

2. DATA PROCESSING AND INTEGRATION

In this section, we describe the detailed approaches to the data integration. The work flow from data processing to final integrated data is shown in Figure 1 and is described in the following subsections.

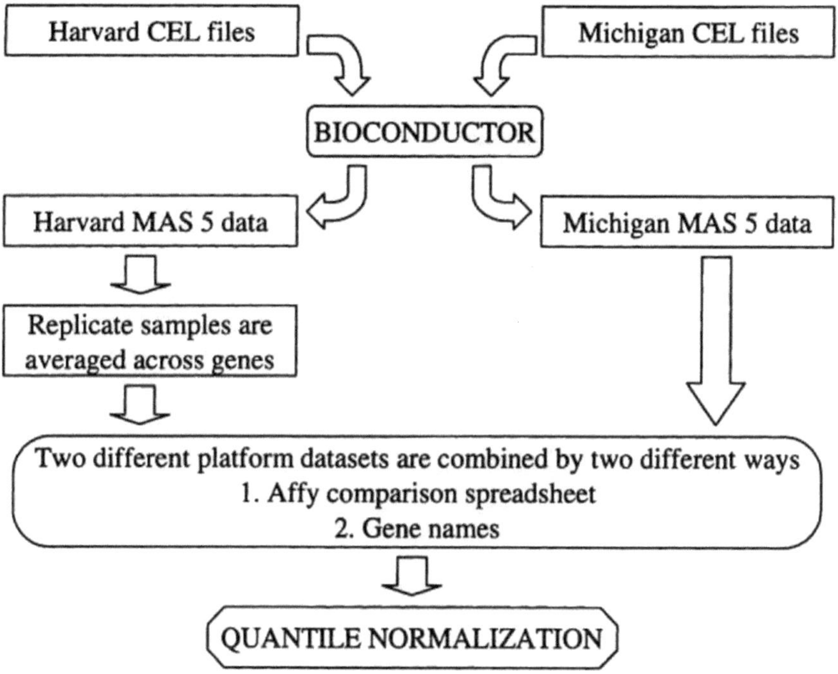

Figure 1. Flow chart for preprocessing of the data.

2.1 Processed Data vs. Raw Data

Processed data from Harvard and Michigan are available but it is possible that they could have been generated by version 4 of Affymetrix MAS. It is well known that MAS 4.0 has many shortcomings compared to the newer version, MAS 5.0 [Affymetrix, 2001]. Those shortcomings include negative expression measures and large variability for genes with low expression values. We have compared the existing Harvard data (processed) with our MAS 5.0 generated data using PCA (principal component analysis) projection. The MAS 5.0 data had better separation of disease groups compared to the existing processed data (Figure 2). Clearer separation of normal lungs and adenocarcinomas from the rest can be seen from MAS 5.0.

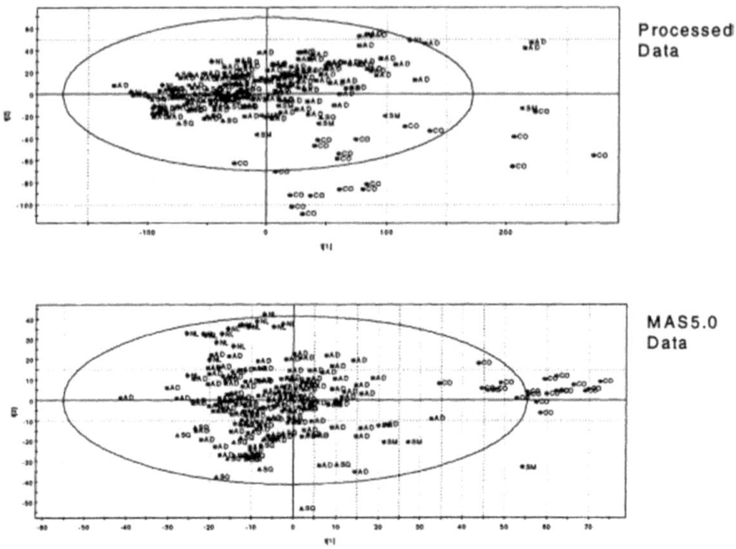

Figure 2. Harvard processed data (top) and MAS5.0 data generated from CEL files (bottom): AD =lug adenocarcinomas and other adenocarcinomas, CO =pulmonary carcinoids, SM =SCLC cases, SQ = aquamous cell carcinomas, and NL= normal.

2.2 Data Integration

We start with the probe level .CEL files. MAS 5.0 expression level data are created by using the affy package [Gautier et al., 2003] in BioConductor (www.bioconductor.org). This summarizes probe level data from each of the Harvard and Michigan data sets. Fifty-one sample pairs in the Harvard data are replicates. For such cases, we average the data across the replicates, and as a result, 254 samples are reduced to 203 samples. However none of the 96 samples in the Michigan data set are replicates.

To integrate the MAS 5.0 data from Harvard and Michigan, we use two different merging methods. As the two experiments used different chips, one-to-one matching is not possible. The first approach matches the probe set ID from the Harvard data (U95a) with the probe set ID from the Michigan data (HuGene FL) using the Array Comparison Spreadsheets (ACS) obtained from Affymetrix homepage (www.Affymetrics.com). We then select those probe sets common to both the Harvard and Michigan data. The second approach integrates the data sets using gene names. The housekeeping genes are not used in this approach. The gene names are obtained from the Affymetrix website. Although probe set names are unique at this point, different probe sets could in some cases correspond to the same gene. When this is true, we average expression levels across the probe sets. The two data sets are then merged by gene name, and we select those genes common to both the Harvard and Michigan data sets.

There are 203 samples (127 lung adenocarcinomas, 12 other adenocarcinomas, 20 pulmonary carcinoids, six SCLC cases, 21 aquamous cell carcinomas, and 17 normal samples) and 12,600 probe sets in the Harvard data set. For the Michigan data, there are 96 samples (86 lung adenocarcinomas and 10 normal samples) and 7,129 probe sets. The integrated dataset combined by using ACS has 6,041 probe sets, while the dataset combined by using gene names, which has 4,837 genes. As mentioned in the preprocessing step, different probe sets could in some cases correspond to the same gene.

We perform PCA to see differences between the two data merging approaches and find that both methods result in approximately the same information. Hence, the integrated data set using gene names will be used for the following analysis.

2.3 Quantile Normalization of Combined Data from Different Platforms

In the final part of the integration, it is necessary to make data from different platforms comparable. Current application of normalization focuses on making the expression distribution of each array comparable. We modify the quantile normalization to make samples comparable across different platforms for each gene. The modification aims to remove differences due to different platforms. The algorithm for the modified quantile-normalization (Q-normalization) is given below.

Q-Normalization Algorithm:

a) Denote $\mathbf{X}=(\mathbf{X}^1, \ldots, \mathbf{X}^k)$, where $\mathbf{X}^m$ represents data from the mth platform with g genes and n_m subjects, $m=1, \ldots k$.
b) Rank each row of $\mathbf{X}^m$ to give $\mathbf{X}^m_{rank}$, $m=1, \ldots k$.
c) Calculate $P^m(i,j)=(X^m\text{rank}(i,j)-1)/(n_m-1)$ where $i=1,..,g$ and $j=1,\ldots,n_m$ for each platform, $m=1,\ldots,k$.
d) For the sth platform, derive $Q^{m,s}(i,j)=P^m(i,j)$-quantile of the ith row of $\mathbf{X}^s$, $s=1,\ldots k$, and $m=1, \ldots k$.
e) The quantile-normalized value $Q^m(i,j)$ of $X^m(i,j)$ is the average of $Q^{m,1}(i,j), \ldots, Q^{m,k}(i,j)$, $m=1, \ldots k$.

Without proper normalization, the integrated Harvard and Michigan data are shown to be completely separated (Figure 3), which means the distributions of the two data sets are not comparable. Figure 4 shows that after Q-normalization, the Harvard and Michigan samples cover the approximately the same projected space by PCA.

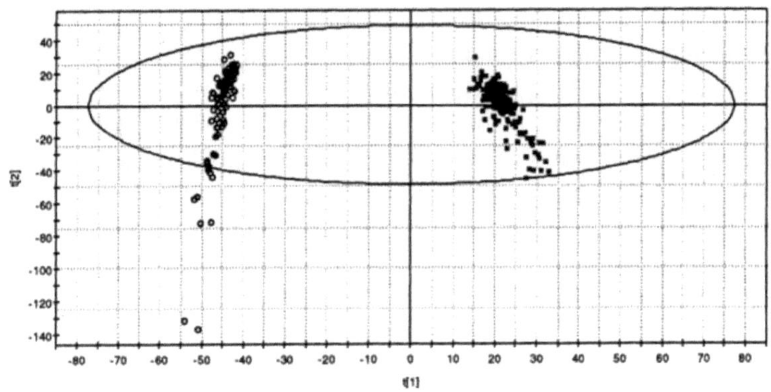

Figure 3. PCA plot of combined data before Q-normalization. The Harvard data are plotted in empty circles and the Michigan in filled squares.

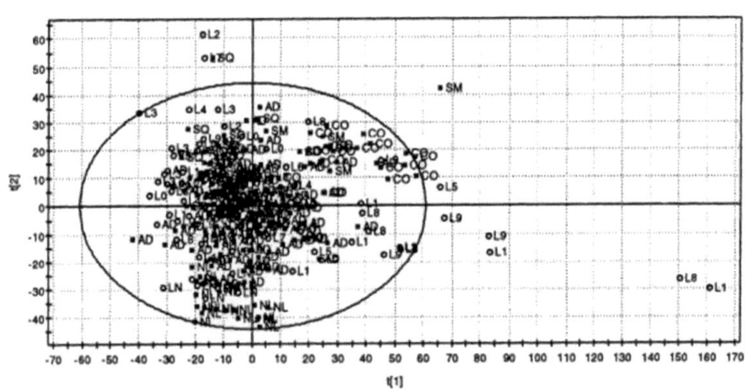

Figure 4. PCA plot of combined data after Q-normalization. The Harvard data are plotted in empty circles and the Michigan data in filled squares.

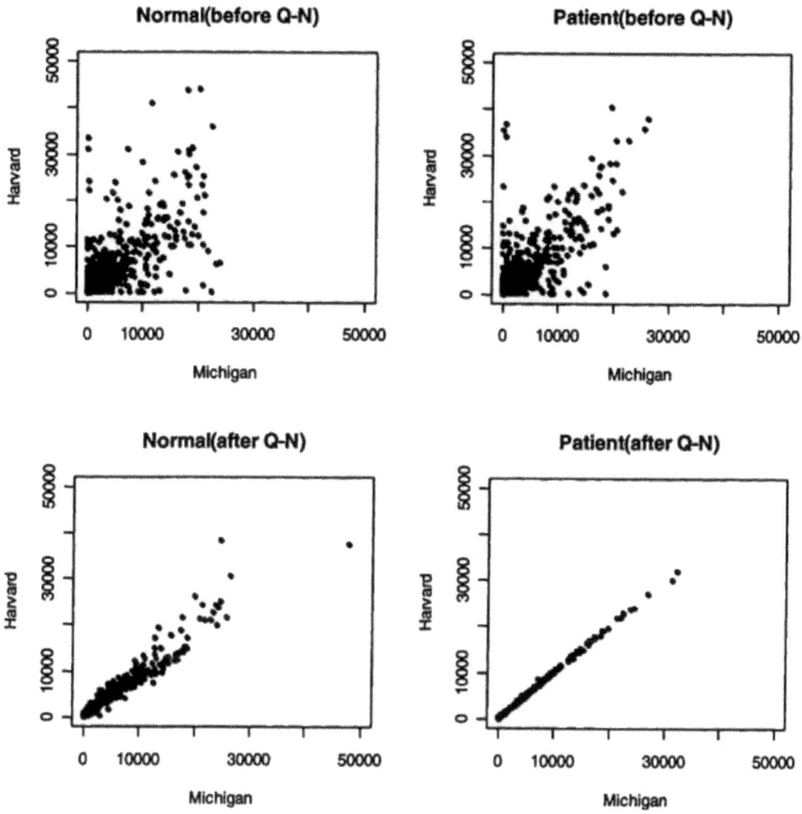

Figure 5. Plot of mean expression values of the Harvard data versus the Michigan data.

Figure 5 shows the scatter plot of the mean expression measures of the Harvard data versus those of the Michigan data before (two plots in the tops) and after (two plots in the bottom) Q-normalization. Each point in the scatter plot corresponds to a gene. The scatter plots confirm that the Q-normalization has made the two data sets comparable with approximately the same distribution. The left two plots are for normal samples while the right two plots are for cancer samples.

3. DISCRIMINATION OF NORMAL LUNGS FROM THE CARCINOMAS

One interesting property of the final data (MAS 5.0, integrated and normalized) is the clear separation of normal lung samples from the rest of the carcinomas. The PCA scores plot in Figure 6 shows the separation of normal lung samples (empty circle) from the adenocarcinomas samples (filled square). One supervised learning projection method is partial least squares (PLS) and its related discriminant analysis (PLS-DA). Figure 7 shows the projection of PLS-DA results. Again normal lung samples have clearly separated themselves from the rest. Using PLS-DA, we can select genes that are responsible for the discrimination of the two classes. Top 20 genes in the Table 1 below is obtained from ranking the genes by their absolute value of PLS-DA regression coefficients.

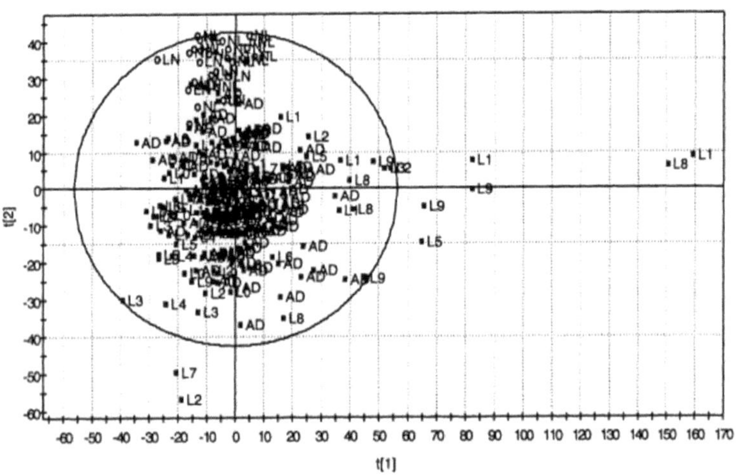

Figure 6. PCA plot of the normal lungs (empty circles) and the adenocarcinomas samples (filled squares).

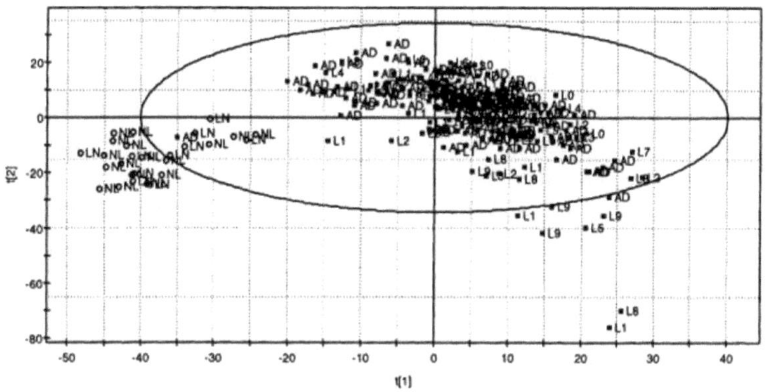

Figure 7. PLS-DA plot of the normal lungs (empty circles) and the adenocarcinomas samples (filled squares).

Table 1. Top 20 genes from ranking of absolute value of PLS-DA regression coefficients.

Gene	Coefficient	Gene	Coefficient
AGER	-0.004889	FABP4	-0.004192
TNA	-0.004784	PTPRB	-0.004157
FHL1	-0.004753	X123	-0.004099
CAV1	-0.004726	GATA2	-0.004098
GPRK5	-0.004576	FMO2	-0.004044
EMP2	-0.004514	GPC3	-0.004033
TNNC1	-0.004476	HYAL2	-0.004022
CA4	-0.004467	FOXF1	-0.003996
PECAM1	-0.004422	CDH5	-0.003963
CLDN5	-0.004254	DF	-0.003912

4. PREDICTION OF ONE PLATFORM FROM ANOTHER

Further validation of integrated data can be done by building predictive models using one of the four lung cancer dataset and then validating those models with one or more of the remaining data sets. Specifically our objective is to build a predictive model to classify lung tissue samples into adenocarcinomas (AD) and normal lung (NL) based on the Harvard data alone and then validate this model by classifying new cases from the Michigan data. Three well-known classification tools, CART, C5 and neural networks (NN), are used to build the predictive models. CART (Classification and Regression Trees) and C5 are two widely used tree-based methods and Neural Networks (NN) is a machine learning method capable of modeling complex, nonlinear functions using a structure consisting of layers of interconnecting nodes or neurons.

We take the integrated data with 4,837 common gene names from the Harvard and Michigan studies. We delete all the observations from tumors other than AD and then create a new binary column (our dependent variable) named AD that indicates whether a tissue sample histologically corresponds to adenocarcinoma (1 = AD, 0 = NL). Then we split the data set into a training data set containing 156 samples (139 AD and 17 NL) from the Harvard study and a test data set containing 96 samples (86 AD and 10 NL) from the Michigan study.

The high dimensionality of our data set (4,837 independent variables) makes it very difficult for the NN to handle. Therefore we have applied a feature selection algorithm based on CHAID (Chi-squared Automatic Interaction Detector) to the training data, to initially select the 50 best predictors. To be consistent, the same 50 predictors are used in all three models to classify the 96 samples in the test data. The performance of each model is summarized in Table 2.

All the models show very high sensitivity (98.84 -100%) but variable specificity (80 – 90%) for classifying new cases. The best performing model is NN with 100% sensitivity and 90% specificity, for an overall accuracy of 98.96%. Both classification tree models (CART and C5) obtain similar results with 98.84% sensitivity and 80% specificity for an overall accuracy of 96.88%.

Table 2. **Summary of performance for the three predictive models on the test data. PPV is the positive predictive value and NPV is the negative predictive value.**

	C5	CART	NN
Sensitivity	98.84%	98.84%	100.00%
Specificity	80.00%	80.00%	90.00%
PPV	97.70%	97.70%	98.85%
NPV	88.89%	88.89%	100.00%
Accuracy	96.88%	96.88%	98.96%

5. SURVIVAL ANALYSIS

Here we consider the time to death as the dependent variable for prediction and use survival analysis to identify those genes associated with high risk of mortality. We use a total of 211 patients (125 from the Harvard data and 86 from the Michigan data) that have both lung adenocarcinoma cancer and survival information.

Since the samples came from two totally independent studies, some study specific factors (known or unknown) might contribute to the risk of mortality. Consequently we need to consider the effect of the different studies in the model when we examine the gene effects. For each gene, we use a frailty (mixed effects) Cox proportional hazard model [Therneau and Grambsch, 2000] with gene as a fixed effect and study effect (Harvard vs. Michigan) as random. Clinical factors might also be used in the model although we do not include them in our model. Genes with significant FDR (false discovery rate) adjusted p-value (at 0.05 level) are listed in Table 3.

Table 3. **Gene list with FDR adjusted p-value less than 0.05.**

Gene Name	Coefficient	Raw p-Value	FDR adjusted p-Value
KIAA0211	-0.0025069	0.000001	0.0054
CTSL	0.0002727	0.000037	0.0313
KRT18	0.0001400	0.000049	0.0313
LHX1	0.0019983	0.000036	0.0313
PGK1	0.0001655	0.000043	0.0313
PRKCBP1	0.0034964	0.000028	0.0313
STX1A	0.0009447	0.000052	0.0313
VEGFC	0.0026009	0.000031	0.0313
P4HA1	0.0010053	0.000065	0.0347
INHA	0.0009011	0.000104	0.0484
RALA	0.0025610	0.000110	0.0484

The above model examines one gene at a time. To examine multiple genes simultaneously, the survival tree method [Therneau and Atkinson, 1997] is used. For efficiency, the genes are first screened using the raw p-value from the Cox model above, resulting in 480 genes at 0.05 level. The survival tree results are shown in Figure 8 which displays the number of samples and the predicted time-adjusted relative event rate (RR, time-adjusted and relative to the whole data set) defined by each node. Based on the tree results, the samples are grouped into high risk (oval) and low risk (hexagon) groups. The Kaplan-Meier plot by risk group is shown in Figure 9. We can see that the mortality behavior is quite different between the two groups.

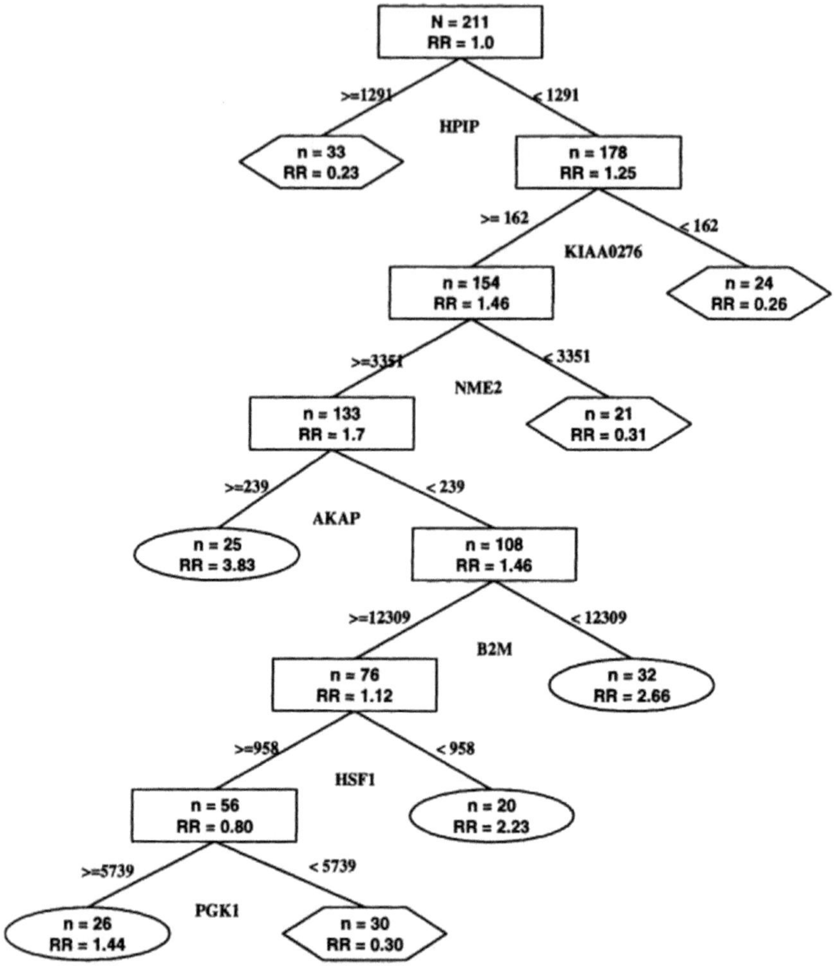

Figure 8. Tree diagram for the results from survival tree method.

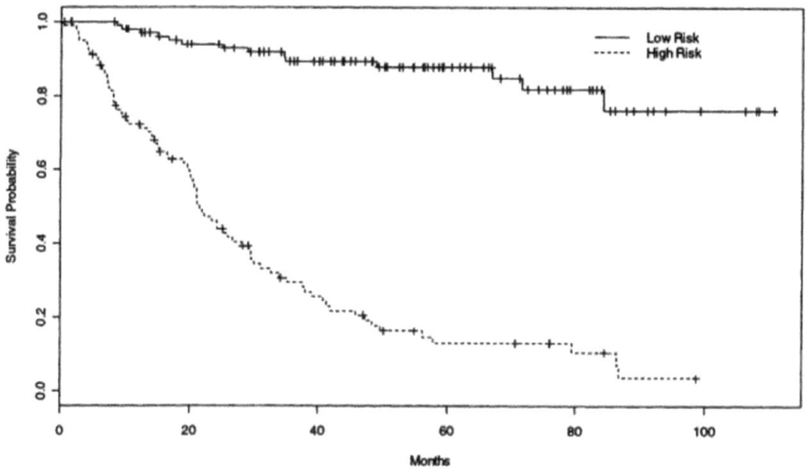

Figure 9. Kaplan-Meier plot of the two risk groups based on the predicted relative risk. There are 108 subjects in the low risk group and 103 in the high risk group. Marked points (+) indicate censored subjects.

6. DISCUSSION AND CONCLUSION

We have demonstrated some effective approaches to overcoming challenges encountered in integrating multiple platform data. These challenges include data processing, data merging, normalization, and data validation. Statistical and data mining methods for survival data were used to analyze the integrated data and it has been shown that we can use the integrated gene expression data to classify the adenocarcinoma samples into different mortality risk groups. Results obtained from this survival analysis would be of biological interest and need further investigation.

For illustratitive purposes we used only two of the four platform data sets. The approaches described in this paper can be generalized to two or more platforms and applied to complicated applications such as integrating data from clinical blood chemistry, gene expression, protein, lipid, NMR, etc. targeted at the same disease. Integrated analysis can provide new insight into the biology of the particular disease by combining information measured from different angles.

The quantile normalization assumes that subjects in different studies have similar characteristics. Otherwise, further modification may be required to take into account of the characteristic differences.

Combining data sets from several studies would provide more samples and give more statistical power in the analysis. However, due to differences in the design of probe sets for different Affymetrix chips, useful information may be lost when we use the combined data sets to do the analysis. For example, only 6,041 probe sets from the Harvard data were kept in the integrated data. About half of the total probe sets in the Harvard data was not used. One possible way to maximize the information would be to integrate the results from the combined data with the results from individual data sets. We are planning to look at the individual data sets in the future and compare the results with the published information.

ACKNOWLEDGEMENTS

Our thanks to Phil Burstein and Alan Menius from GlaxoSmithKline for their encouragement and support for doing research on the integration and analysis of biological data. We also like to thank our colleague Keith Crowland for his careful review of the paper which helped make it more readable.

REFERENCES

Affymetrix (2001) Affymetrix Microarray Suite User Guide. Version 5 edition, Affymetrix, Santa Clara, CA.

Beer D, Kardia S, Huang C, Giordano T, Levin A, Misek D, Lin L, Chen G, Gharib T, Thomas D, Lizyness M, Kuick R, Hayasaka S, Taylor J, lannettoni M, Orringer M and Hanash S (2002) Gene-expression profiles predict survival of patients with lung adenocarcinoma. Nature medicine, Vol 8, 816-824

Bhattacharjee A, Richards, WG, Staunton J, Li C, Monti S, Vasa P, Ladd C, Beheshti J, Bueno R, Gillette M, Loda M, Weber G, Mark EJ, Lander ES, Wong W, Johnson BE, Golub TR, Sugarbaker DJ, and Meyerson M. (2001) Classification of human lung carcinomas by mRNA expression profiling reveals distinct adenocarcinomas subclasses. PNAS, Vol 98, 13790-13795

Bolstad BM (2001) Probe level quantile normalization of high density oligonucleotide array data. Technical report of Division of Biostatistics, University of Berkeley.

Gautier L, Cope L, Bolstad BM and Irizarry RA (2004) Textual description of affy. http://www.bioconductor.org.

Therneau TM and Atkinson EJ (1997) An introduction to recursive partitioning using the RPART routines. Technical Report #61, Division of Biostatistics, Mayo Clinic.

Therneau TM and Grambsch PM (2000) *Modeling Survival Data,* Springer, New York.

Chapter 7

ENTROPY AND SURVIVAL-BASED WEIGHTS TO COMBINE AFFYMETRIX ARRAY TYPES AND ANALYZE DIFFERENTIAL EXPRESSION AND SURVIVAL

[1]JIANHUA HU, [2]GUOSHENG YIN, [2]JEFFREY S. MORRIS, [2]LI ZHANG, [1]FRED A. WRIGHT
[1]The University of North Carolina at Chapel Hill, Chapel Hill, NC 27599; [2]The University of Texas, MD Anderson Cancer Center, Houston, TX 77030

Abstract: In order to comprehensively identify genes with expression levels that correlate with survival for patients with lung adenocarcinoma, we combined data across the Harvard and Michigan studies. Two different versions of Affymetrix oligonucleotide microarrays were used in these two studies. We proposed combining arrays of different platforms by assigning weights to the expression levels of each gene across data sets based on the entropy of the residual matrix. In each data set, the expression level of each gene is quantified by the "reduced" model proposed by Li and Wong [2001], which is equivalent to a method using the singular value decomposition. We combined information across different chip types by first identifying common genes on the two chip types, and then assigning weights based on residual entropy for each gene. To incorporate clinical information, especially survival data, in detecting important genes, we proposed a new method based on weighted t-tests (wtt). The survival information can be absorbed into a set of weights assigned to the expression intensities across all the arrays or subjects, based on the predicted median survival time using the Cox proportional hazards model. Important genes can be identified by comparing the survival-weighted t-tests with another t-test comparing the cancer patients to the reference group, and error rates can be controlled by permutation procedures.

Key words: Entropy; false discovery rate; median survival time; SVD; weighted t-test

1. INTRODUCTION

DNA microarray technology has been increasingly used and is beginning to play an important role in many areas of biomedical research. This technology allows us to monitor the expression levels of very large numbers of genes simultaneously and repeatedly in cell lines, human tissues and a wide range of organisms. The two popular types of platforms are the spotted cDNA microarrays and oligonucleotide arrays.

The distinctive feature of the oligonucleotide array technology is the effective utilization of multiple probes. Multiple oligonucleotides of different sequences are hybridized onto different regions of the same RNA that are complementary to the oligonucleotides. The other source of redundancy is the use of mismatch (MM) probes, which are each identical to a corresponding perfect match (PM) probe, except for a single base that is mutated at the central position (typically the 13th position). The design of oligonucleotide arrays with PM/MM probe sets can help to distinguish whether a detected signal is real or only a chance artifact due to nonspecific cross-hybridization or other measurement errors. It may have improved differentiating ability compared to that of the cDNA array, which uses a single spot probe.

The two studies with oligonucleotide array data are chosen for our exploration. The Michigan data set [Beer et al., 2002] uses the HU6800 platform with 20 probe pairs, which produces 7,129 probe sets. The Harvard [Bhattacharjee et al., 2001] data set uses a different and newer type of platform, the U95A, which contains 12,625 probe sets with 16 probe pairs each. Issues arising in microarray experiments include the preprocessing of raw data, normalization techniques, and the experimental design. A recent tendency in research is to incorporate other information such as sequence data and clinical data into the analysis of gene expression data. Here, the Harvard and Michigan studies follow in this trend. One of their common objectives was to identify important genes that are related to lung adenocarcinoma, a disease that is the leading cause of cancer deaths in the United States. The survival data of patients in the studies were used together with the gene expression profiles to achieve this goal.

Our research objective was to further combine information from the two different studies, and identify genes that have significant impact on primary lung adenocarcinomas. We focused on the non-cancer samples and on the histologically-defined lung adenocarcinoma samples, since they represent the most common histology and are accompanied by relatively complete survival data. We proposed a novel method to combine the two gene expression data sets based on the singular value decomposition (SVD) method and "entropy". Moreover, we proposed a new weighted t-test for incorporating the clinical information into the procedure of identifying important genes.

2. EXAMINING SURVIVAL AND GENE EXPRESSION DATA

2.1 Survival Data

To examine the homogeneity of the two populations across the Harvard and Michigan studies, we started by comparing the clinical variables, including survival data. Patient data from the two studies had comparable distributions of age, sex, and smoking status. However, only tumors of stages 1 and 3 were represented in the Michigan study, while tumors of stages 1, 2, 3 and 4 were represented in the Harvard data. We dichotomized the stage variable by combining the local stages (1 and 2) and the advanced stages (3 and 4). Figure 1 contains the Kaplan-Meier curves for the two studies. There is a significant difference in survival between the two studies based on the log-rank test (p-value = 0.01). We included an indicator variable to account for an institution effect in the analysis, and otherwise the populations seemed comparable for a common pooled analysis

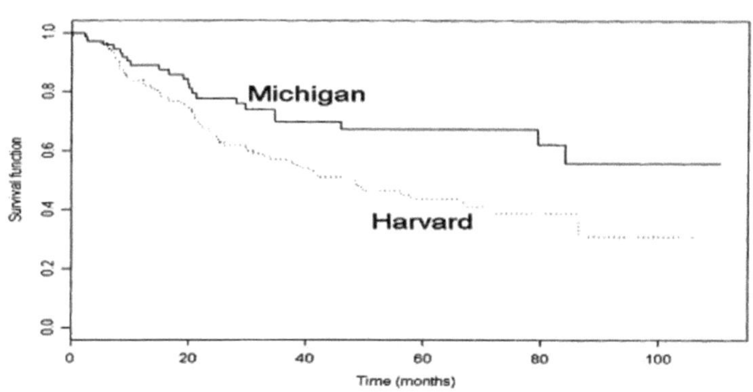

Figure 1. Kaplan-Meier plots for Harvard and Michigan studies.

2.2 Gene Expression Data

We log-transformed raw intensities of gene expression for each array and plotted them to remove bad chips. Samples L54, L88, L89, and L90 in the Michigan arrays contained a large round dark spot at the center of the chip (see Figure 2), and samples L22, L30, L99, L81, L100, and L102 contained a large number of extremely bright outliers according to MAS5.0 (Affymetrix, Inc.). Two outlier chips were detected and removed in the Harvard dataset using dChip (CL2001040304 and CL2001041716). We kept the most recently dated run among the Harvard samples with 48 replicate arrays (the arrays had been duplicated due to a bad first run).

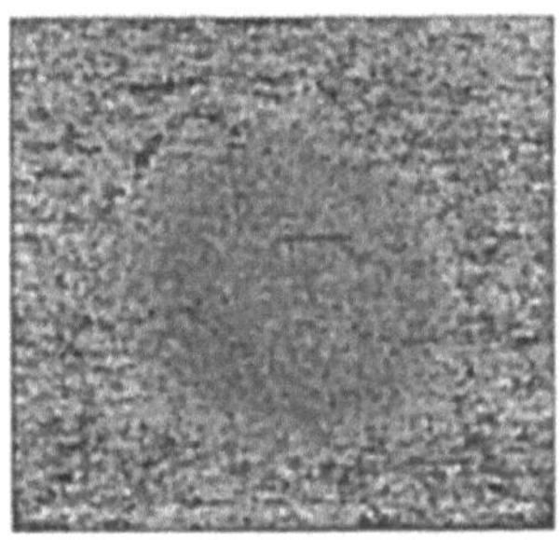

Figure 2. Image plot of log-expression for sample L88 in Michigan data set. Green and red indicate log-expression levels below and above the median for the chip, indicating a bad chip. Samples L54, L88, L89, and L90 have similar plots.

This preprocessing resulted in a data set with matching clinical and microarray data for 229 patients, which includes control and primary lung adenocarcinomas samples, 143 from Harvard with 17 references, and 86 from Michigan with 10 references.

3. NORMALIZATION AND EXPRESSION INDEX ESTIMATION

Microarray normalization is an important issue. It is a process to remove the unwanted variation in microarray experiments that affects the measured gene expression levels. Because scanned images may have a different level of overall brightness, it is important to normalize arrays such that they have comparable levels of brightness before analyzing gene expression levels. Because the model-based expression index analysis involves different arrays simultaneously, the comparable brightness of the arrays needs to be assured.

Although still an active research area, this issue has been extensively discussed and explored in the literature.

Li and Wong [2001] developed an iterative procedure to determine an "invariant set", namely, the set of non-differentially expressed genes. Keeping the array that has the median overall brightness (the baseline array) as the invariant one, all the other arrays are normalized to it. Instead of using a real array as the baseline array, we computed the median expression intensity across all the arrays for each gene and defined the reference array as the collection of all the median intensities. Then, we normalized all the arrays to it by fitting the usual linear regression models.

The term "expression index" describes a statistic used to represent an expression level for a particular gene that is estimated from raw hybridization intensities on the array. Estimation of the expression index becomes an important issue because all the statistical tests and inferences are made based on the indices. In recent years, various statistical methods for modeling the gene expression levels have been proposed, including nonparametric approaches and parametric models. A multiplicative model proposed by Li and Wong [2001] is feasible and popular with biologists, and is advantageous because of higher efficiency of the estimates than others. We performed the Li-Wong reduced model (LWR) using the SVD technique [Hu et al., 2003], because of the direct connection between the two techniques. The first characteristic mode (see Holter et al., 2000 for definition) of the data matrix for each gene, i.e. $\mathbf{PM}_{I\times J}\text{-}\mathbf{MM}_{I\times J}$, is proportional to the corresponding LWR estimates, where I and J denote the numbers of arrays and probes, respectively. The new method is more efficiently and closely related to the method of combining different platforms that is described below.

4. COMBINING DATA FROM DIFFERENT AFFYMETRIX ARRAYS

Determining how to combine the different types of Affymetrix oligonucleotide chips in the two studies was one of our main challenges. The Hu_FL Affymetrix chip with 20 probe pairs was used in the Michigan study, while the newer version HG_U95aV2 chip with 16 probe pairs was used by the Harvard group. A list of common probe sets representing the same gene between these two different chip types is available at the following dChip URL:

http://www.biostat.harvard.edu/complab/dchip/info_file.htm#common_probeset_file.

There are 5,987 probe set pairs representing the same genes across the two studies. However, due to differences in probe densities and probe sequences, the expression levels of the genes in these two chip types are not directly comparable. In order to obtain comparable gene expression levels across the two chip types, we introduced a technique for assigning weights to each expression index in the two data sets.

An important concept involved in our approach is entropy [Shannon, 1948]. The entropy $H(f)$ of an absolutely continuous density $f(x)$ is defined $H(f) = \int f(x) \log f(x) dx$. H can be viewed as a measure of randomness or unpredictability of a random variable X, and has been applied in a variety of hypothesis testing problems. Some papers, e.g., Vasicek [1976] and Dudewicz and Meulen [1981] discussed the construction of hypothesis tests on normality or uniformity based on [0,1] using this concept. Another important application of entropy is in combination with SVD for genome-wide expression data [Alter et al., 2000]. Using ideas from Alter et al. [2000], we defined "fraction of eigenintensity" as $p_j = \frac{\sigma_j^2}{\sum_{j=1}^{J} \sigma_j^2}$, where J is the number of probes and σ_j denotes the ith eigenvalue from the SVD decomposition. It indicates the degree of structure in the data matrix that can be captured by the ith eigenvector for arrays and probes. The discrete analogue of the Shannon entropy of a given data set is

$$e = \frac{-1}{\log(J)} \sum_{j=1}^{J} p_j \log(p_j) \tag{1}$$

where the entropy is scaled so that $0 \leq e \leq 1$. e describes the "randomness" of the data matrix, in the sense that SVD cannot meaningfully discern structure in fitting the data. In particular, $e=0$ corresponds to an ordered and redundant data set where all the expression is captured by a single eigenvalue, and $e=1$ corresponds to a disordered and random data set.

Assuming that the LWR is the true model from which the underlying expression index can be estimated, there should be no systematic pattern left in the residual matrix after fitting the model. This procedure is equivalent to subtracting the product of the first set of eigenvalues of the data matrix and two eigenvectors from it using the SVD. The randomness of the residual matrix can be assessed by the distribution of its eigenvalues, quantified by the entropy. We reasoned that the data that better fit the model should have a higher entropy. First, in each study, the expression intensity matrix of each gene was standardized to a mean of 0 and a variance of 1 (to avoid one source of bias in the SVD). After applying the SVD, we obtained the eigenvalue entropies of the residual matrices for each gene. The distribution of the entropies across all the common genes in each data set is shown in Figure 3. Overall, the Harvard data appears much better, with entropies centered around 0.9, while those from Michigan are widely spread from 0 to 1. However, a few genes from the Harvard study were assigned very low weights (some even close to 0).

The two studies have different dimensions in their data matrices for each gene. However, this fact has little impact on the entropies of the residual matrix, as demonstrated in some limited simulations. For each gene, the two entropy values (Harvard and Michigan) were then standardized to make them sum up to 1, and then within each study the appropriate weight was multiplied by the expression index to obtain a new entropy-weighted expression index. The weight is proportional to the entropy value, with a larger weight being assigned to the model-based expression index estimate in the study that has higher entropy for the specific gene.

To assess the performance of the entropy weighting strategy, we used the false discovery rate (FDR) as a comparison criterion. The FDR is defined as the expected proportion of false rejections (truly null) among the rejected

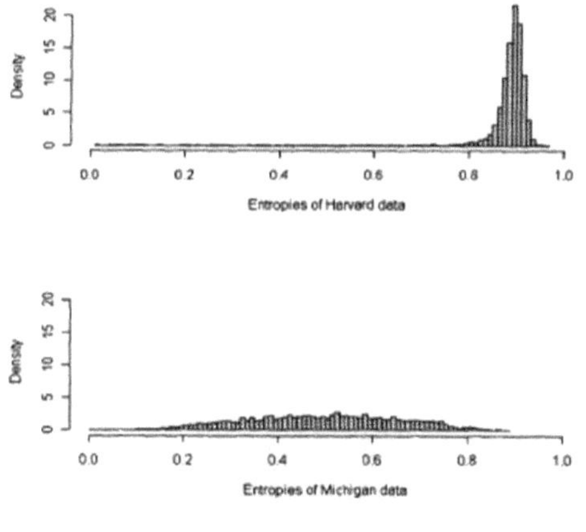

Figure 3. Distributions of entropies in the Harvard and Michigan studies.

hypotheses [Benjamini and Hochberg, 1995]. We followed the permutation procedures as implemented in the software SAM [Tusher et al., 2001] to estimate the FDR. We computed ordinary t statistics based on both the unweighted and weighted expression data, and also conducted 5000 permutations of the t-tests. In each permutation, we randomly drew 27 samples from the total of 229 patients to be treated as the "reference" and treated the rest as the "cancer" patients. We estimated the FDR as a function of the number of genes that can be detected. Figure 4 shows the relationship between the FDR and the number of rejected genes up to 300. Clearly, the weighted data yielded a dramatically lower FDR level than the unweighted one.

Moreover, we examined the correlation among the samples. Ideally, the correlations within the reference or disease samples should be higher than between the reference and disease samples, if indeed gene expression can be used to discriminate between the groups. Examining the within-reference and within-disease samples in each data, we found the weighted method can increase the correlations over the unweighted. We calculated the differences of the pairwise correlations between the weighted and unweighted expression data, where 26.5% differences are between 0.1 and 0.5. The rest of the correlation differences vary around 0. We also compared the correlations between the reference and disease groups across the two data sets, and found that 78.8% pairwise correlations are lower in the weighted expression data, though the differences were not dramatic.

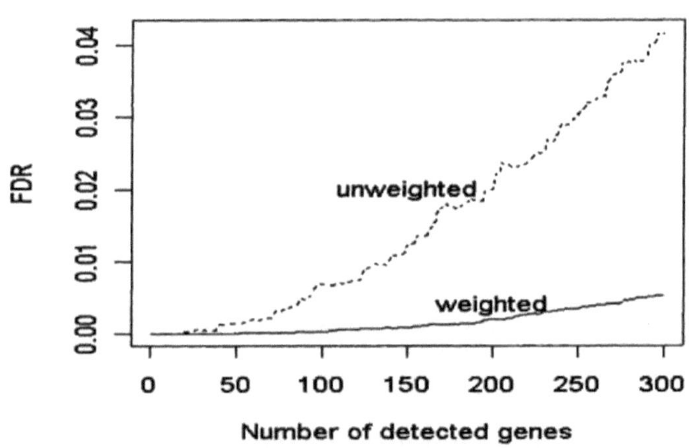

Figure 4. Comparison of FDR between weighted and unweighted expression data.

5. IDENTIFYING IMPORTANT GENES

5.1 Weighting Based on Survival Data

Another major goal in this analysis is to combine the gene expression data with the patient survival data. To find those genes that either directly affect or can help predict patient survival, we needed to take into account clinical information other than survival time and censoring, such as, tumor stage, age, sex, and smoking status. Clearly, the Cox proportional hazards model is readily applicable. However, adding both the clinical variables and the gene data into the Cox model may cause a high-dimensionality problem. To address this issue, we introduced a new method of the weighted t-test (wtt). To incorporate the clinical information, some form of weight needed to be constructed for gene expression intensity data. Due to censoring, the construction of appropriate weights for each subject was quite challenging. In order to obtain reasonable weights, we proposed using the predicted median survival time, as described below.

A total of 229 subjects were in the pooled sample, 188 of which were cancer patients with available recorded survival information. Our analysis included the institution, age, sex, smoking status, and tumor stage as covariates. The institution was examined because the survival curves were very different between the studies performed at Harvard and at Michigan. For the *i*th subject with a covariate vector Z_i, the Cox proportional hazards model is given by

$$\lambda(t \mid Z_i) = \lambda_0(t)\exp(\beta^T Z_i), \tag{2}$$

where $\lambda_0(t)$ is the unknown and unspecified baseline hazard function and β is the regression parameter of interest. For the *i*th subject, the survival function is given by

$$S(t \mid Z_i) = \exp\{-\Lambda_0(t)\exp(\beta^T Z_i)\}, \tag{3}$$

where $\Lambda_0(t)$ is the cumulative baseline hazard function. The parameter estimates are listed in Table 1.

We constructed the predicted survival curve for each subject based on the clinical information only, from which we estimated the median survival time: $m_i = \inf\{t : S(t \mid Z_i) < 0.5\}$.

Table 1. Parameter estimates under the Cox proportional hazards model (H.R. is the hazard ratio and S.E. is the standard error).

Covariate	Regresson Coefficient	H.R.	S.E.	p-value
Institution	0.6392	1.89	0.2501	0.011
Age	0.0267	1.03	0.0120	0.027
Sex	0.1292	1.14	0.2288	0.570
Smoking Status	0.0063	1.01	0.0032	0.048
Tumor Stage	1.5552	4.74	0.2666	<0.001

We assigned an averaged median survival time to those subjects with missing survival information. For a given subject, regardless of whether an observation is a failure or is censored, m_i is determined by the covariate Z_i, which circumvents the potential bias caused by the censored data. Based on the predicted median survival time, we calculated the weights that were proportional to m_i, for each cancer patient accordingly,

$$w_i = \frac{m_i}{\sum_{i=1}^{n} m_i} \times n \tag{4}$$

Moreover, for subjects in the reference sample, we did not assign any weights because they were controls and were all alive at the end of the study. Thus, we computed the weights using all the common clinical variables provided in the two studies.

With the survival-weighted expression data, we conducted a two-sample t-test for each gene to measure the difference in expression levels between the control group and the cancer patients. We also performed t-tests on the expression data with no weight adjustment. In order to find the genes related to survival information, we examined the difference between the t-test statistics after and before the survival-weight adjustment, i.e., $d_k=t_{after}-t_{before}$, for the kth gene, k=1 ...,5,987. Note that by substracting the t_{before} values, d_k was constructed to be sensitive to effects of expression on survival, and not on mere differences in expression in cancer vs. reference. We again performed 5,000 permutations. In each permuted dataset, we implemented the ordinary t-test and survival wtt, and recorded their statistics together with d_k for each gene. We ordered d_k for each permutation, and let $d_{(k)}$ denote the ordered d_k. Then, we calculated the averaged order statistics, $\bar{d}_{(k)}$, across all the 5,000 permutations. A gene was deemed to be related to survival when $d_{(k)}-\bar{d}_{(k)}$ (if $d_{(k)}$ was positive) was larger than an appropriate threshold, or when $d_{(k)}-\bar{d}_{(k)}$ (if $d_{(k)}$ was negative) was smaller than some threshold. We thus obtained a list of the most significant genes.

To accommodate the multiple testing issue in our analysis, again we applied the FDR criterion and identified the 12 genes most significantly related to survival as described above, while controlling for the FDR at 0.05. Furthermore, the statistical significance of the detected genes could be measured by p-values obtained from the permutation procedure, defined as the proportion of the difference d_k at least as extreme as that observed. A list of the 12 genes is shown in Table 2, along with the names of probe sets in U95a and Hu6800 platforms, gene descriptions and corresponding p-values. Here, as with Table 3 described below, we found an intriguing number of sex-specific genes and hormones. Sex was specifically included in the Cox model, and these results suggested taking a closer look at expression of these genes within each sex, and also for any possible lack of proportional hazards across the two sexes. Several other genes, including ribosomal proteins and those involved in immune response and cell differentiation, are typical of broad functional characteristics that have appeared in other cancer studies.

Table 2. List of 12 most important genes related to survival, ordered from the most to the least significant to the least using SAM.

U95A	Hu6800	Gene Annotation	p-value
725_i_at	J03071_cds3_f	Chorionic Somatomammotropin Hormone Cs-5	<0.001
35281_at	U31201_cds1	laminin, gamma 2 (nicein, kalinin, BM600, Herlitz junctional epidermolysis bullosa)	<0.001
34961_at	M88282	T cell activation, increased late expression	<0.001
31838_at	U79274	protein predicted by clone 23733	<0.001
37174_at	D14660	mitochondrial ribosomal protein L19	0.035
530_at	U16258	ribosomal protein S7	0.005
41643_at	X83301_s	cluster includes X83301:H.sapiens SMA5 mRNA/cds=(319,741)/gb=X83301 /gi=603029	<0.001
35894_at	X14362	complement component (3b/4b) receptor 1, including Knops blood group system	<0.001
32864_at	L10102_rna1	sex determining region Y	0.002
32686_at	D86096_cds6	prostaglandin E receptor 3 (subtype EP3)	<0.001
722_at	D87957	rcd1 (required for cell differentiation, Spombe) homolog 1	0.009
1338_s_at	X13930_f	X13930 /FEATURE=cds Human CYP2A4 mRNA for P-450 IIA4 protein	0.008

5.2 Differentiating between reference and cancer subjects

The wtt method can be used to identify the important genes that are differentially expressed in the two groups of reference and cancer patients. We had more confidence in choosing for further biological validation the genes found to have significant results under both tests. The rationale is that such genes show both a difference between cancer vs. reference and also have an apparent effect on survival. Again, we used the SAM-like

Table 3. The 15 most significant genes differentiating between reference and cancer groups, ordered from the most to the least significant according to the sum of ranks using SAM.

U95A	Hu6800	Gene Annotation	rank(|d(k)-d(k)|) in t-test	rank(|d(k)-d(k)|) in wtt
725_i_at	HG1751-HT1768	Chorionic Somatomammotropin	1	1
33780_at	M36200	vesicle-associated membrane protein 1 (synaptobrevin 1)	5	2
40081_at	HG3945-HT4215	phospholipid transfer protein	3	8
220_r_at	S76756_s	S76756 4R-MAP2=microtubule-associated protein, isoform	12	4
35281_at	U31201_cds1	laminin, gamma 2	2	15
38150_at	U22233	methylthioadenosine phosphorylase	8	10
32461_f_at	HG3137-HT3313	zinc finger protein 81 (HFZ20)	13	6
37263_at	U55206	gamma-glutamyl hydrolase (conjugase, folylpolygammagl-h)	11	13
37975_at	X04011	cytochrome b-245, beta polypeptide (granulomatous disease)	17	9
37399_at	D17793	aldo-keto reductase family 1, member C3 (3-alpha h-d)	21	7
36287_at	X83368	phosphoinositide-3-kinase, catalytic, gamma polypeptide	18	11
1197_at	D00654	D00654 / DEFINITION=HUM ACTSG7 Homo sapiens gene	25	5
1482_g_at	L23808	matrix metalloproteinase 12 (macrophage elastase)	27	3
36617_at	HG3342-HT3519_s	inhibitor of DNA binding 1, dominant negative h-l-h protein	20	12

U95A	Hu6800	Gene Annotation	rank(|d(k)-đ(k)|) in t-test	rank(|d(k)-đ(k)|) in wtt
35462_at	U17033	phospholipase A2 receptor 1, 180kD	10	32

procedure to identify significant positive genes. By sorting $|d_{(k)}-\bar{d}_{(k)}|$ and taking into consideration the sign of $d_{(k)}$ in both the ordinary t-test and wtt, the 15 most significant genes with the smallest sums of ranks of $|d_{(k)}-\bar{d}_{(k)}|$ across the two t-test statistics were identified. Table 3 shows the names of the 15 probe sets in U95a and Hu6800 platforms, gene annotations and the ranks using the two different statistics.

6. CONCLUSIONS

In this study, we conducted the expression data analysis by using LWR estimates as the underlying gene expression index estimates. We implemented LWR based on the SVD method due to its efficiency and consistency with the method that we proposed for combining different array types. We imposed a SVD entropy weight on the expression of each gene, thereby demonstrably achieving a lower FDR level in comparison of cancer vs. reference samples. The approach of using residual entropy to judge the quality of expression estimates can be applied in a much more general context. We incorporated survival data by imposing another weighting scheme based on the predicted median survival time to each subject. To identify important genes having significant impact on patient survival, we compared a survival weighted t-test to the corresponding ordinary t-test, with both tests using the entropy-weighted combined expression values. We assessed the significance test of the difference between the weighted and unweighted t statistics by permutation procedures. Moreover, based on the two t-tests, we identified those genes that were differentially expressed between the reference and cancer groups. Clearly, the proposed method can be extended to more general situations, for instance, to an F-test in the case of dealing with multiple samples. And the power property of the survival weighted t-test method needs to be explored. Regarding normalization procedures, certainly nonlinear or nonparametric models are more flexible and may fit better. This will be explored in our future research.

7. ACKNOWLEDGEMENTS

We thank Kevin Coombes and Fei Zou for helpful discussions on this project.

8. REFERENCES

Alter, O., Brown, P. O. and Botstein, D. Singular value decomposition for genome-wide expression data processing and modeling. *PNAS,* 97: 10101-10106, 2000

Beer, D, Kardia, S. L. R., Huang, C, Giordano, T. J., Levin, A. M., Misek, D. E., Lin, L., Chen, G., Gharib, T. G., Thomas, D. G., Lizyness, M. L., Kuick, R., Hayasaka, S., Taylor, J. M. G., lannettoni, M. D., Orringer, M. B. and Hanash, S. et al. Gene-expression profiles predict survival of patients with lung adenocarcinoma. *Nature Medicine,* 9: 816-824, 2002.

Benjamini, Y. and Hochberg, Y. Controlling the false discovery rate: a practical and powerful approach to multiple testing. *Journal of the Royal Statistical Society Series B,* 57: 289-300, 1995.

Bhattacharjee, A, Richards, W. G., Staunton, J., Li, C, Monti, S., Vasa, P., Ladd, C., Beheshti, J., Bueno, R., Gillette, M., Loda, M., Weber G., Mark, E. J., Lander, E. S., Wong, W., Johnson, B. E., Golub, T. R., Sugarbaker, D. J. and Meyerson, M. Classification of human lung carcinomas by mRNA expression profiling reveals distinct adenocarcinoma subclasses. PNAS, 98: 13790-13795, 2001.

Cox, D. R. Regression models and life tables (with discussion). *Journal of Royal Statistical Society, Series B,* 34: 187-220, 1972.

Dudewicz, E. J. and Meulen, E. C. V. D. Entropy-based tests of uniformity. *Journal of the American Statistical Association,* 76: 967-974, 1981.

Holter, N. S., Mitra, M., Maritan, A., Cieplak, M., Banavar, J. R. and Fedoroff, N. V. Fundamental patterns underlying gene expression profiles: simplicity from complexity. *PNAS,* 97: 8409-8414, 2000.

Hu, J., Wright, F. A. and Zou, F. An adaptive SVD approach of estimating expression Indexes for oligonuleotide arrays. *Manuscript,* 2003.

Lemon, W. J., Palatini, J. J., Krahe, R. and Wright, F. A. Theoretical and experimental comparisons of gene expression indexes for oligonucleotide arrays. *Bioinformatics,* 18: 1470-1476, 2002

Li, C. and Wong, W. H. Model-based analysis of oligonucleotide arrays: expression index computation and outlier detection. *PNAS,* 98: 31-36, 2001.

Shannon, C. E. A mathematical theory of communication. Bell System Technical Journal 27: 379-423, 623-656, 1948. Reprinted in *Key Papers in the Development of Information Theory* (1974), ed. D. Slepian, New York: IEEE press, 5-2.

Tusher, V. G., Tibshirani, R. and Chu, G. Significance analysis of microarrays applied to the ionizing radiation response. *PNAS,* 98: 5116-5121, 2001.

Vasicek, O. A test for normality based on sample entropy. *Journal of the Royal Statistical Society Series B,* 38: 54-59, 1976.

Chapter 8

ASSOCIATING MICROARRAY DATA WITH A SURVIVAL ENDPOINT

Sin-Ho Jung, Kouros Owzar, Stephen George

Affiliation: Department of Biostatistics and Bioinformatics, Duke University Medical CenterDurham, North Carolina

Abstract: In many microarray studies the primary objective is to identify, from a large panel of genes, those which are prognostic markers of a censored survival endpoint such as time to disease recurrence or death. These genes are considered prognostic in that their respective expressions are associated, in an appropriate sense, with the survival endpoint of interest. From a practical point of view, this requires not only specifying a appropriate measure of association and a suitable statistic thereof, but also, as the number of genes is large, proper handling of the consequential issue of multiplicity. In this paper, we will address the aforementioned issues by utilizing a general correlation measure and a non-parametric statistic, and by controlling the family-wise error rate by employing permutation resampling. Comprehensive simulation studies are conducted to investigate the statistical properties of the proposed procedure. The proposed procedure is demonstrated with microarray data.

Key words: Censoring, family-wise error rate, rank correlation, multiple testing

1. INTRODUCTION

In early microarray studies, for example [Golub et al., 1999], the primary objective focused on identifying genes which express differentially in different phenotypes. More recently the objectives have expanded to include discovering the relationship between gene expression level and aggressiveness of a disease (such as cancer) or the existence of tumor residue after tumor resection. The most popular and often useful endpoint in this type of study may be time to a clinical event, such as disease recurrence or death. In this context, a gene is considered to be prognostic if its expression level is associated with the survival endpoint. The times to such events are usually subject to censoring due to loss to follow up or termination of the study.

When considering or devising a statistical method for analysis of such studies, the following issues need to be taken into account. Firstly, one needs to choose a measure of association which properly quantifies the dependence between the survival endpoint and each of a large number of genes. Secondly, one needs to specify a statistic which robustly estimates this measure of association for each gene. Finally, given that the number of genes under consideration is large, it is imperative to ensure that the overall error-rate is, in some appropriate sense, adequately controlled.

A simple heuristic approach to this end is to partition the subjects into two groups: event versus no event, and proceed by using a standard approach, such as a two-sample t-test, to identify genes differentially expressing between the two groups (see for example André et al. [2002] and Shannon et al. [2002]). This approach, however, can be biased as the subjects in the study usually have different follow-up periods and some patients may not have had enough follow-up period to observe events.

Park et al. [2002] and Nguyen et al. [2002] reduce the dimension of gene expression data using a method like principal component analysis and fit a Cox's regression model using the derived components as covariates. However, this approach fails to test on the marginal correlation of a gene (or a principal component) and the survival variable and fails to adjust for the multiplicity of the testing procedure.

Dhanasekaran et al. [2001] identified a prognostic gene based on p-values calculated by fitting a Cox's regression model without adjusting for multiplicity of the original genes. Wigle et al. [2002] fit an univariate Cox regression model on each gene expression level and applied the approach discussed in Dubey [1993] to the resulting univariate (or unadjusted) p-values to adjust for the multiple testing procedure. Sørlie et al. [2001] also fit univariate Cox's regression models on gene expression levels and applied a method called SAM (Significance Analysis of Microarrays), as for example discussed in Tusher and Tibshirani [2001], to discover prognostic genes.

For each gene, Jenssen et al. [2002] sort the expression level observations and partition all patients into two groups using each order statistic as a cutoff: one group for those patients who have gene expression levels smaller than the cutoff and the other for those who have gene expression levels equal to or larger than the cutoff. The (standardized) logrank statistic is calculated to compare the survival distribution between the two groups. They take the largest logrank statistic with respect to all possible cutoffs for each gene and apply a Bonferroni correction to identify prognostic genes adjusting for multiple testing. They argue that the choice of maximum logrank test statistics yields an anti-conservative procedure, but the conservative Bonferroni adjustment works in the opposite direction. This method does not provide an accurate control of the family-wise error rate (FWER).

In this paper, we use a measure of rank correlation between a continuous variable and a survival variable. This measure was originally proposed by O'Quigley and Prentice [1991] and was subsequently used by Jung et al. [1995] to compare two correlated surrogate markers which are prognostic for patient's survival time. We use this rank correlation measure to associate each gene expression level with a survival variable, and discover prognostic genes using a single-step multiple testing method outlined by Jung [2003], which uses a permutation method to derive adjusted p-values for the genes. Simulation studies are conducted to evaluate the performance of the proposed procedure. To demonstrate the applicability of the procedure to real microarray data a case study is presented.

2. MULTIPLE TESTING USING A RANK CORRELATION

First, we describe a rank correlation between the expression level of a gene (a continuous variable) and a survival endpoint. Suppose that there are n subjects. For patient i, T_i denotes the time to an event (such as tumor recurrence or death), called survival time hereafter. The survival time may be censored due to loss to follow-up or study completion, so that we observe $X_i = \min(T_i, C_i)$ together with censoring indicator $\Delta_i = I(T_i \le C_i)$, where C_i is the censoring time which is assumed to be independent of T_i given gene expression level. Let $Y_i(t) = I(X_i \geq t)$ and $N_i(t) = \Delta_i I(X_i \leq t)$ be the at-risk process, that takes 1 if the subject i is at risk at time t and 0 otherwise, and the death process, that takes 1 if the subject i has an event at or before t and 0 otherwise , respectively. Let $Y(t) = \sum_{i=1}^{n} Y_i(t)$.

Let m denote the number of genes (or multiple tests) under consideration and $(Z_{ij}, 1 \leq j \leq m)$ denote the expression levels of the m genes from patient i. Usually the gene expression data within each subject are correlated.

As a general measure of association between the expression level for gene j and the survival data, we use

$$W_j = \sum_{i=1}^{n} \int_0^{\infty} \left(R_{ij} - \frac{\sum_{i'=1}^{n} R_{i'j} Y_{i'}(t)}{Y(t)} \right) dN_i(t)$$

$$= \sum_{i=1}^{n} \Delta_i \left(R_{ij} - \frac{\sum_{i'=1}^{n} R_{i'j} I(X_{i'} \geq X_i)}{\sum_{i'=1}^{n} I(X_{i'} \geq Y_i)} \right), \qquad (1)$$

where R_{ij} is the rank of Z_{ij} among $(Z_{1j}, ..., Z_{nj})$. Note that W_j has a form of covariance between R_{ij} and death process $N_i(t)$. W takes a large positive (negative) value if the gene tends to overexpress in the high (low) risk patients, and distribute around 0 if the gene expression does not have any impact on the survival.

W is rank-invariant with respect to Z as well as T. Furthermore, W is the same as the score test based on Cox's partial likelihood for a proportional hazards model in which the rank of Z_i is used as a time-independent covariate (see [O'Quigley and Prentice 1991]).

Jung et al. [1995] used this measure to compare two correlated markers ('genes' here) which are prognostic for survival time. Contrary to O'Quigley and Prentice [1991], we do not assume any (semi-)parametric model between survival and gene expression level in this paper.

We want to identify genes that are associated with survival time. We consider hypotheses,

$$H_j : T \text{ and } Z_j \text{ are not associated,} \tag{2}$$

versus

$$\bar{H}_j : T \text{ and } Z_j \text{ are negatively associated,} \tag{3}$$

i.e., gene j tends to overexpress in high-risk patients. Then, we may reject H_j in favor of $\bar{H}_j$ for a large value of W_j. Let $H_0 = \cap_{j=1}^{m} H_j$, under which no genes are associated with survival time. Given FWER α, we want to find a common critical value c_α that satisfies

$$P\{\cup_{j=1,\dots,m}(W_j \geq c_\alpha) \mid H_0\} = P(\max_{j=1,\dots,m} W_j \geq c_\alpha \mid H_0) \leq \alpha. \tag{4}$$

In order to solve (4), we need to know the joint distribution of $(Z_1,\dots,Z_m)$ under H_0. However usually this is not available in a closed form, especially due to the extremely high dimension of the random vector. So, we propose to use a permutation method to approximate the null distribution of the test statistics.

In order to maintain the correlation structure among m genes, we keep the m gene expressions $(Z_{i1},\dots,Z_{im})$ together. We generate permutation data under H_0 by separating the survival data (X_i,Δ_i) from the gene expression data $(Z_{i1},\dots,Z_{im})$, and randomly matching the survival data with the gene expression data. For a permutation $(j_1,\dots,j_n)$ of $(1,\dots,n)$, a permutation sample is generated as $\{(X_{j_i},\Delta_{j_i},Z_{i1},\dots,Z_{im}), i=1,\dots,n\}$. Since our test statistics depend on the gene expression data only through their ranks, we may replace the gene expression data with their ranks, i.e. a permutation sample is given as $\{(X_{j_i},\Delta_{j_i},R_{i1},\dots,R_{im}), i=1,\dots,n\}$.

From the b-th permutation sample, we calculate the test statistics $w_1^{(b)},\dots,w_m^{(b)}$ and $\bar{w}^{(b)} = \max_{j=1}^{m} w_j^{(b)}$. The number of possible permutations, $n!$, as for example for a moderate sample size $n = 10$, we have $n! = 3{,}628{,}800$, is typically rather large. We may choose a reasonably large number of these permutations, say B = 10,000. Then, from (1), c_α is approximated by the $[B(1-\alpha)+1]$-st order statistic of $\bar{w}^{(1)},\dots,\bar{w}^{(B)}$, where [a] is the largest integer that is smaller than a.

An adjusted p-value for gene j is defined as the minimum FWER at which H_j will be rejected. So, with an observed test statistic value $W_j = w_j$ for gene j, the adjusted p-value is given as

$$p_j = P(\max_{j'=1,\ldots,m} W_{j'} \geq w_j \mid H_0), \tag{5}$$

which can be estimated from the permutations:

$$p_j \approx \frac{\sum_{b=1}^{B} I(\overline{w}^{(b)} \leq w_j)}{B}. \tag{6}$$

Jung [2003] investigated a similar testing procedure for multiple two-sample t-tests.

If we want to identify the genes either positively or negatively associated with survival time, then we may use two-sided tests. For marginal two-sided tests, we want to find a common critical value $\tilde{c}_\alpha$ that satisfies

$$P(\max_{j=1,\ldots,m} |W_j| \geq \tilde{c}_\alpha \mid H_0) \leq \alpha. \tag{7}$$

We can approximate $\tilde{c}_\alpha$ using the same permutation method described above except that we obtain

$$\overline{w}^{(b)} = \max_{j=1,\ldots,m} |W_j^{(b)}| \tag{8}$$

from the b-th permutation data. Adjusted p-value for gene j, with observed test statistic $W_j = w_j$, also should be modified as

$$p_j = P(\max_{j'=1,\ldots,m} |W_{j'}| \geq |w_j| \mid H_0), \tag{9}$$

which is approximated as

$$p_j \approx \frac{\sum_{b}^{B} I(\overline{w}^{(b)} \leq |w_j|)}{B}. \tag{10}$$

Given FWER α, we may reject H_j if $W_j > c_\alpha$ or $p_j < \alpha$. Calculation of c_α involves sorting of $(\overline{w}^{(b)}, 1 \leq b \leq B)$, so that the testing procedure using adjusted p-values requires less computing time.

3. NUMERICAL STUDIES

We investigate the performance of the proposed single-step multiple testing procedure with a large number of genes, m. We generate gene expression data from a multivariate normal distribution and survival time from a lognormal distribution, which is negatively correlated with prognostic genes. In type I error analyses, we generate the data as follows. For iid N(0,1) random numbers $\tau_i, \varepsilon_{i0}, \varepsilon_{i1}, \ldots, \varepsilon_{im}$, we set

$$\log(T_i) = \tau_i$$
$$Z_{ij} = \varepsilon_{ij}\sqrt{1-\rho} + \varepsilon_{i0}\sqrt{\rho} \text{ for } 1 \le j \le m. \quad (11)$$

Then, the survival time is not associated with any genes, and the gene expression data have a multivariate normal distribution with zero means, unit variances and a compound symmetric correlation structure with coefficient ρ. We consider m = 1, 000, n = 20 or 50, $\rho = 0$, .3 or .6, and 20% or 40% censoring. A censoring time is generated from $U(0, c_0)$ with c_0 chosen for 40% censoring. With c_0 fixed at this value, a censoring variable for 20% censoring is generated from $U(c_1, c_0 + c_1)$ by choosing a proper c_1 value. Null distribution of the test statistic is approximated from B = 1,000 random samples of $n!$ possible permutations. Empirical FWER is computed as the proportion of samples rejecting H_0 by our testing procedure with one-sided FWER=.05 among N = 1,000 simulations. Simulation results are reported in Table 1. Our procedure overall has an empirical FWER close to the nominal level.

Table 1. Empirical FWER for nominal 5% FWER with m=1,000, B=1,000 and N=1,000.

	n=20			*n*=50		
Censoring	ρ=0	.3	.6	ρ=0	.3	.6
20%	.055	.054	.052	.053	.048	.045
40%	.044	.035	.046	.053	.057	.054

For power analyses, the first D genes are set to be prognostic with correlation coefficients r with log(T). The data are generated as follows. For iid N(0,1) random numbers $\tau_{i0}, \tau_i, \varepsilon_{i0}, \varepsilon_{i1}, \ldots, \varepsilon_{im}$, we obtain

$$\log(T_i) = \tau_i\sqrt{1-r} - \tau_{i0}\sqrt{r} \quad (12)$$

and

$$Z_{ij} = \begin{cases} \varepsilon_{ij}\sqrt{1-\rho} + \varepsilon_0\sqrt{\rho} + \tau_{i0}\sqrt{r} & \text{for } 1 \le j \le D \\ \varepsilon_{ij}\sqrt{1-\rho} + \varepsilon_0\sqrt{\rho} & \text{for } D+1 \le j \le m \end{cases} \tag{13}$$

It can be shown that $\mathrm{corr}(\log T_i, Z_{ij}) = -r/\sqrt{1+r} \equiv \eta$ for $1 \le j \le D$ and $= 0$ for $D+1 \le j \le m$; $\mathrm{corr}(Z_{ij}, Z_{ij'}) = (\rho + r)/(1+r)$ for $1 \le j < j' \le D$, $= \rho/\sqrt{1+r}$ for $1 \le j \le D < j' \le m$ and $= \rho$ for $D+1 \le j < j' \le m$. Note that η is the parameter of interest. We set $n = 50$, $D = 5$, 10 or 15; $\eta = .3$ or .6 in addition to the parameters set for the type I error analyses. The simulation results are summarized in Table 2.

Table 2. Empirical rejection rate of each H_j under n=50, m=1,000, B=1,000 and N=1,000. Genes are grouped for prognostic ones (j=1,...,D) and non-prognostic ones (j=D+1,...,m). The numbers in parentheses are empirical rejection rate of any of these hypotheses, called global power.

η	D	Censoring	Genes	ρ=0	.3	.6
.3	5	20%	$j\le D$	.001-.006	.001-.003	.002-.009
			$j>D$	.000-.002	.000-.002	.000-.003
				(.076)	(.062)	(.071)
		40%	$j\le D$	.000-.003	.001-.004	.003-.012
			$j>D$	.000-.002	.000-.002	.000-.004
				(.049)	(.063)	(.079)
	15	20%	$j\le D$	.000-.006	.000-.006	.003-.009
			$j>D$	.000-.002	.000-.002	.000-.003
				(.084)	(.081)	(.092)
		40%	$j\le D$	.000-.003	.001-.007	.003-.012
			$j>D$	.000-.002	.000-.003	.000-.004
				(.061)	(.077)	(.090)
.6	5	20%	$j\le D$	.042-.059	.051-.071	.098-.120
			$j>D$	.000-.001	.000-.002	.000-.003
				(.259)	(.268)	(.307)
		40%	$j\le D$	.024-.043	.035-.048	.069-.090
			$j>D$	.000-.002	.000-.002	.000-.003
				(.186)	(.199)	(.228)
	15	20%	$j\le D$	.041-.066	.051-.072	.097-.129
			$j>D$	.000-.001	.000-.002	.000-.003
				(.502)	(.448)	(.459)
		40%	$j\le D$	.023-.044	.030-.050	.070-.096
			$j>D$	.000-.002	.000-.002	.000-.004
				(.343)	(.322)	(.347)

As illustrated in Table 2, for non-prognostic genes the false rejection rates, i.e. the probability that H_j is rejected when H_j is true, are very low. Global power, i.e. the probability that any H_j is rejected, and true rejection rate, i.e. the probability that H_j is rejected when $\bar{H}_j$ is true, increase in η. With $\eta = .3$, global power and true rejection rate are low. But with $\eta = .6$, global power and true rejection are very high. True rejection rate increases in ρ, but global power does not seem to change in ρ.

Beer et al. [2002] used oligonucleotide arrays to generate gene expression data for $m = 4966$ genes from $n = 86$ patients with lung adenocarcinoma. We applied our multiple testing method to their data to identify prognostic genes. Analysis results are summarized in Table 3.

Table 3. Analysis results for Michigan Data (n=86,m=4966) with B=10,000 permutations. Genes with at least one adjusted one-sided p-value smaller than .8 are listed. (- meaning a one-sided adjusted p-value of 1.0000)

	Adjusted p-value			Unadjusted p-value		
	ρ<0	ρ>0	ρ≠0	ρ<0	ρ>0	ρ≠0
S1P	–	.0125	.0227	–	.0000	.0000
KIAA0153	.7131	–	.8794	.0006	–	.0009
KIAA0263	–	.7714	.9145	–	.0004	.0011
NULL	–	.6767	.8490	–	.0003	.0006
NP	.0426	–	.0769	.0001	–	.0001
SLC2A1	.5555	–	.7504	.0003	–	.0005
STX1A	.1976	–	.3229	.0000	–	.0002
GPC3	–	.7423	.8961	–	.0007	.0010
TMSB4X	–	.3387	.5101	–	.0004	.0004
SELP	–	.5421	.7314	–	.0000	.0003
VEGF	.6509	–	.8300	.0002	–	.0005
FUCA1	–	.7022	.8687	–	.0007	.0010
PRKACB	–	.2588	.4099	–	.0000	.0002
HPIP	–	.4986	.6919	–	.0001	.0003
SERPINB5	.3894	–	.5720	.0001	–	.0002
FUT3	.6065	–	.7926	.0004	–	.0010
NUCB1	–	.7530	.9043	–	.0007	.0009
P2RX5	–	.4731	.6666	–	.0002	.0006
NULL	.3148	–	.4847	.0001	–	.0004
MS4A2	–	.6863	.8571	–	.0004	.0008
GRO3	.3781	–	.5570	.0000	–	.0000
SORT1	–	*.2987*	*.4593*	–	*.0000*	*.0000*

In Table 3, the columns with $\rho < 0$ ($\rho > 0$) are for testing the one-sided alternative hypotheses that a gene tends to overexpress in high (low) risk patients. The columns $\rho \neq 0$ are for two-sided tests. Adjusted and unadjusted p-values are listed for those genes with either one-sided adjusted p-value smaller than 0.8. We observe that Gene SIP underexpresses and Gene NP overexpresses in high risk patients. For all other genes listed in Table 3, the unadjusted p-values are very small. The corresponding adjusted p-values for these genes, however, are not small enough for statistical significance after adjusting for multiplicity of the testing procedure.

4. CONCLUSIONS

This paper presents a comprehensive non-parametric procedure for analyzing microarray studies whose primary outcome measure is censored survival time. For a method to be useful in these types of microarray data analysis, it must address the following three issues:

a) The ability to quantify the degree of association and the corresponding statistical significance between *each* gene and the survival variable.
b) The ability to control the *overall* error rate.
c) Robustness against outliers and model misspecification.

As illustrated in the literature review presented in the introductory section, there is a sizable literature on analyzing microarray studies whose primary endpoint is a censored survival variable. What the proposed method attempts to accomplish is to address *simultaneously* all of the three aforementioned issues. Furthermore, as this method is inferential, rather than data-driven, it will not only be useful from the point of view of exploratory data analysis, but should also serve as an invaluable tool for sample size and power calculations in designing experiments for which microarray studies with survival endpoints are planned.

To demonstrate the performance as well as applicability of the method, we have presented simulation as well as case studies. The simulation studies suggest that the false-rejection rate (i.e., incorrectly declaring a non-prognostic gene as prognostic) for this method is virtually negligible. For moderately sized studies (e.g., n = 50), the method will have very good global power (i.e., probability of detecting at least one of the prognostic genes) as long as the hypothesized effect size is reasonably large (e.g.,

$\eta = 0.6$). Also, in such cases, the method enjoys good true-discovery rates (i.e., correctly declaring a prognostic gene as prognostic). Furthermore, the method adequately controls the FWER.

The amount of association between the survival endpoint and the expression level of gene j was quantified estimated by W_j. One can generate variations of the proposed method by employing other types of association measures and statistics. Such extensions are subject to active pursuit by the authors.

5. REFERENCES

André, A., Karn, T., Solbach, C, Seiter, T., Strebhardt, K., Holtrich, U., and Kaufmann, M., 2002, Identification of high risk breast-cancer patients by gene expression profiling. Lancet, 359, 131-132.

Beer, D. G., Kardia, S. L., Huang, C. C., Giordano, T. J., Levin, A. M., Misek, D. E., Lin, L., Chen, G., Gharib, T. G., Thomas, D. G., Lizyness, M. L., Kuick, R., Hayasaka, S., Taylor, J. M., Iannettoni, M. D., Orringer, M. B., and Hanash, S., 2002, Gene-expression profiles predict survival of patients with lung adenocarcinoma. Nature medicine, 8, 816-824.

Dhanasekaran, S. M., Barrette, T. R., Ghosh, D., Shah, R., Varambally, S., Kurachi, K., Pienta, K. J., Rubin, M. A., and Chinnaiyan, A. M., 2001, Delineation of prognostic biomarkers in prostate cancer. Nature, 412(6849), 822-6.

Dubey, S. D., 1993, Adjustment of p-values for multiplicities of intercorrelating symptoms. Pages 513-527 of: Statistics in the pharmaceutical industry (second edition).

Golub, T. R., Slonim, D. K., Tamayo, P., Huard, C., Gaasenbeek, M., Mesirov, J. P., Coller, H., Loh, M. L., Downing, J. R., Caligiuri, M. A., Bloomfield, C. D., and S., Lander E., 1999, Molecular classification of cancer: Class discovery and class prediction by gene expression monitoring. Science, 286(15), 531-537.

Jenssen, T. K., Kuo, W. P., Stokke, T., and Hovig, E., 2002, Associations between gene expressions in breast cancer and patient survival. Hum genet, 111, 411-20.

Jung, S. H., 2003, Single step multiple testing, submitted.

Jung, S. H., Wieand, S., and Cha, S. S., 1995, A statistic for comparing two correlated markers which are prognostic for time to an event. Statistics in medicine, 14, 2217-2225.

Nguyen, D. V., and Rocke, D. M., 2002, Tumor classification by partial least squares using microarray gene expression data. Bioinformatics, 18(1), 39-50.

O'Quigley, J., and Prentice, R. L., 1991, Nonparametric tests of association between survival time and continuously measured covariates: The logit-rank and associated procedures. Biometrics, 47, 117-127.

Park, P. J., L., Tian, and S., Kohane I., 2002, Linking gene expression data with patient survival times using partial least squares. Bioinformatics, 18, S120-S127.

Shannon, W. D., Watson, M. A., Perry, A., and Rich, K., 2002, ntel statistics to correlate gene expression levels from microarrays with clinical covariates. Genetic epidemiology, 23, 87-96.

Sørlie, T., Perou, C. M., Tibshirani, R., Aas, T., Geisler, S., Johnsen, H., Hastie, T., Eisen, M. B., van de Rijn, M., S., Jeffrey S., Thorsen, T., Quist, H., O., Matese J. C. Brown P., Botstein, D., Eystein Lonning, P., and L., Borresen-Dale A., 2001, Gene expression patterns of breast carcinomas distinguish tumor subclasses with clinical implications. Pnas, 98(19), 10869-74.

Tusher, V. G., and Tipshirani, R., 2001, Significance analysis of microarrays applied to the ionizing radiation response. Pnas, 98(9), 5116-21.

Wigle, D. A., Jurisica, I., Radulovich, N., Pintilie, M., Rossant, J., Liu, N., Lu, C., Woodgett, J., Seiden, I., Johnston, M., Keshavjee, S., Darling, G., Winton, T., Breitkreutz, B.-J., Jorgenson, P., Tyers, M., Shepherd, F. A., and Tsao, M. S., 2002, Molecular profiling of non-small cell lung cancer and correlation with disease-free survival. Cancer res, 62(11), 3005-3008.

Chapter 9

DIFFERENTIAL CORRELATION DETECTS COMPLEX ASSOCIATIONS BETWEEN GENE EXPRESSION AND CLINICAL OUTCOMES IN LUNG ADENOCARCINOMAS

Kerby Shedden[1] and Jeremy Taylor[2]
[1]Dept. of Statistics, University of Michigan [2]Dept. of Biostatistics, University of Michigan

Abstract: We propose a simple data analysis procedure that aims to uncover an association between gene expression and the status of a clinical outcome variable. Rather than focus on differences in group means, as is usually done, we search for pairs of genes such that the strength or direction of their association is linked to the value of the outcome variable. This more complex pattern of gene expression, which we call "differential correlation", may be especially relevant in studying clinical outcomes such as survival and grade, since it has often been difficult to identify marker genes whose mean expression varies directly with such outcomes. In applying our method to two lung cancer microarray data sets, we discovered that a substantially greater number of genes are likely to be associated with clinical outcomes such as tumor stage via differential correlation than are associated via changes in mean expression.

Key words: Differential correlation, gene expression, interaction

1. INTRODUCTION

Important clinical disease characteristics such as survival times and tumor stage often fail to exhibit strong associations with gene expression. One possible reason for this may be that complex clinical responses are biologically manifested in subtle ways, and hence may not be easily detectable using conventional statistical measures that look for expression shifts in the average levels of single "marker genes".

We propose a simple analysis method for relating gene expression levels to binary response variables that aims to detect a link between the degree of association within a pair of genes and the clinical response. This "differential correlation" is a more subtle form of association compared to differences in mean expression ("differential expression"). Although differential correlation is a more complex statistical measure, from a biological viewpoint it resembles a simple gene regulatory interaction. As such regulation is known to be altered in the progression of many cancers, the idea that differential correlation may occur for cancer-related endpoints is well supported biologically.

The organization of this article is as follows. Section 2 contains a general description of our proposed differential correlation methodology, and Section 3 contains the results of applying the method to two lung cancer gene expression datasets. In Section 4 we discuss some limitations and possible future directions.

2. DIFFERENTIAL CORRELATION

Complex clinical assessments such as survival or tumor stage do not always exhibit clear associations with gene expression. In many cases, the number of genes with significant mean difference between two groups of samples is comparable to the number of significant differences found under randomization. This may be due to low statistical power in the study design, but more fundamentally it may be due to a small or vanishing number of genes whose mean expression level varies directly with the levels of the response variable.

Nevertheless, there may be marked effects on expression covariation related to the clinical outcome. Here we investigate the most simple such effect -- a change in the association between two genes as the outcome varies. In the case of binary outcomes, this suggests identifying pairs of genes whose correlation coefficient differs significantly between the two levels of the response variable.

Ideally, one can propose a mechanistic explanation for any particular instance of differential correlation. For example, a pair of genes whose expression is more tightly correlated for the less severe state of the outcome variable compared to the more severe state may reflect a decoupling of expression associated with disease progression, perhaps resulting from loss of a common regulator. A pair of genes whose co-expression increases with disease progression may reflect genes acting in concert to produce tumor phenotypes such as vascularization or rapid growth.

To identify genes exhibiting differential correlation, for each pair of genes i,j, we calculated the robust correlation coefficient (biweight midcorrelation, Wilcox [1997]) between the expression levels of the two genes within each group. Suppose this yields correlation coefficients ρ_1 and ρ_2. The difference $\Delta_{ij} = \rho_1 - \rho_2$ measures the increase or decrease in correlation between the two groups. We selected genes where $|\Delta_{ij}| > 0.6$ for further analysis. The Δ_{ij} statistic could also be constructed using standard Pearson correlations, but we found that with relatively small sample sizes, large shifts in Pearson correlation were often due to a single outlying sample.

Randomization was used to assess whether all candidate gene pairs exhibiting differential correlation can be explained by random variation. Specifically, the outcome variable levels were uniformly permuted across the samples, and the Δ_{ij} values were recomputed for each pair of genes in the permuted data. If the number of Δ_{ij} values in the actual data exceeding a given threshold (we use 0.6) was greater than the 95th or 99th percentile under randomization, we deemed it likely that at least some of the gene pairs exhibiting differential correlation are biologically significant.

3. ANALYSIS OF THE LUNG TUMORS

3.1 Data Integration

We analyzed data collected using two Affymetrix microarrays. The University of Michigan data (originally reported in Beer et al. [2002]) were obtained using the full length (HuFL) array that has 7,129 probesets. The Harvard data (originally reported in Bhattacharjee et al. [2001]) were obtained using the U95A array that has 12,625 probesets. Our analysis focused on the adenocarcinoma samples, of which there are 79 in the Michigan dataset and 84 in the Harvard dataset (after averaging duplicates and removing samples with low tumor cellularity).

Both datasets were obtained using Affymetrix microarrays, where the perfect match (PM) and mismatch (MM) intensities for a set of probes (a "probeset") carry the expression information for one transcript. We used the trimmed mean of the PM-MM differences (across the probes) as the numerical summary for a probeset. A detailed discussion of data processing can be found online: http://dot.ped.med.umich.edu:2000/pub/index.html.

We mapped each probeset to a Unigene accession number using the array annotation files available from the Affymetrix web site. There were 5,141 distinct Unigene accession numbers that mapped to at least one probeset on both arrays. The majority of the Unigene numbers mapped to a single probeset on each array, but some mapped to as many as eight probesets.

To construct a numerical summary for each Unigene accession number, we averaged the probeset summaries within each sample across the probesets that map to a common Unigene accession number. These averages (henceforth referred to as gene expression levels) were then left-truncated at zero, right-shifted by 50, $\log_2$ transformed, and quantile normalized (for details, see the above reference to our data processing methods). The gene expression levels exhibited reproducible aggregate characteristics between the two datasets -- within-dataset means had correlation 0.52 across genes, and within-dataset standard deviations had correlation 0.56 across genes.

The most variable genes were considered for subsequent analysis. We selected the 1,102 genes having standard deviation greater than 0.5 in both data sets. This is a rather high standard deviation threshold, leaving only highly variable genes. For purposes of illustrating our methodology, we selected genes according to this strict rule, but a more complete biological investigation might use a lower threshold.

Clinical outcome variables that were measured in similar ways in the two studies and that could easily be dichotomized were selected for analysis. These variables were survival (24 months survival vs. death before 24 months, omitting censored cases), stage (I vs. III), grade (well and moderate vs. poor), smoking status (less than 10 pack years vs. 10 or more pack years), and K-Ras mutation status (wild type vs. mutant). Table 1 contains the number of samples at each level, for each outcome variable in the two data sets.

Information from the two datasets was combined by requiring that differential correlation thresholds be met independently in both data sets, with consistent direction of change. That is, for a pair of genes i,j to be considered differentially correlated with respect to a particular outcome variable, it was required that the condition $|\Delta_{ij}| > 0.6$ be met in both datasets, and that the sign of Δ_{ij} be the same in both datasets.

3.2 Baseline Analysis

We began by carrying out a baseline analysis using standard methods for detecting differential mean expression. For each clinical response variable, the samples from each dataset were stratified into two groups, which were subsequently compared at each gene using two sample t-tests and fold changes. For each dataset, three sets of genes were identified: (i) genes having a t-test p-value smaller than 0.05, (ii) genes having a 2-fold or greater change in mean expression, and (iii) genes having a 1.5-fold or greater change in mean expression. Next the genes satisfying (i) for both datasets were selected, and from these only the genes with consistent direction of expression change in the two datasets were retained. Similarly, genes satisfying (ii) or (iii) and having consistent direction of expression change in the two datasets were considered. The numbers of such genes for each outcome are given in Table 2. Next to each observed number are the 95th and 99th percentiles under randomization, estimated from 300 randomizations.

Table 1. Sample sizes for the Michigan and Harvard data sets.

Outcome	Level	U. Mich.	Harvard
Early Death	<=24 months	17	30
	>24 months	60	53
Stage	I	60	62
	III	19	8
Grade	Well/Moderate	58	29
	Poor	20	14
Smoking	<10 pack years	14	12
	>=10 pack years	63	72
K-Ras	Wild type	42	39
	Mutant	37	24

Table 2. Three differential mean expression measures (t-test and two levels of fold change) compared to differential correlation for five clinical outcomes.

Outcome	t-test	2-fold	1.5-fold	Diff. Corr.
Early Death	13(5,16)	0(0,1)	2(5,16)	62(115,165)
Stage	10(6,14)	0(1,3)	9(6,14)	1444(1328,1361)
Grade	75(6,12)	2(0,1)	22(6,12)	920(641,889)
Smoking	19(5,8)	16(5,8)	16(5,8)	98(82,100)
K-Ras	21(5,12)	0(0,0)	6(5,12)	210(190,255)

The results of the baseline analysis (Table 2, columns 2-4) indicated that a small number of genes were differentially expressed for each outcome, except for grade, which produced a moderate level of differential expression. An even smaller number of genes exhibited differences that were large in magnitude. Nevertheless, based on the randomization analysis, the t-test results were statistically significant for all five outcome variables, and it is unlikely that more than half of the identified genes are false positives.

Many of the genes identified in the baseline analysis as being associated with the clinical outcomes do not have known biological functions that are easy to relate to the biological nature of the outcome. However several genes associated with proliferation exhibit significant association with tumor grade. PCNA, Cyclin B1, TOPIIA, and BOP1 are upregulated in poorly differentiated tumors, reflecting the likely faster growth rate of poorly differentiated cancer cells.

3.3 Results of the differential correlation analysis

3.3.1 Randomization analysis and global significance

We identified pairs of genes with differential correlation greater than 0.6 in both datasets, or smaller than -0.6 in both datasets. These pairs were identified from among the $\approx 6\times10^5$ distinct pairs that can be formed from the 1,102 genes meeting the variability conditions. The fifth column of Table 2 shows the results of this analysis. Compared to the randomized results, stage, grade, smoking, and K-Ras show an excess of differentially correlated pairs, while early death does not.

The biological significance of this finding is that it suggests that some of the clinical outcomes have a much broader relationship with gene expression than is indicated by differential mean expression. For example, while only 10 genes exhibit strong evidence of differential mean expression with stage, the 1,444 pairs showing significant differential correlation with stage include 858 distinct genes (60% of all genes considered). While the randomization analysis suggests that many of the 1,444 pairs may be false positives, even if only 100 pairs are truly differentially correlated (taking a very conservative view of the randomization results), these pairs are likely to contain far more than 10 distinct genes.

3.3.2 An example – negative interaction of BENE and Hs. 143288 is specific to poorly differentiated tumors

Focusing now on a specific example, Figures 1 and 2 show a pair of genes that are differentially correlated with respect to grade in both data sets. Figure 1 shows that the genes BENE and Hs. 143288 have little association in well or moderately differentiated samples (perhaps there is a positive trend in the Harvard data, but this is quite weak). On the other hand, Figure 2 shows a strong negative trend between the two genes in the poorly differentiated samples. For both data sets, high levels of Hs. 143288 expression are associated with low levels of BENE expression. Adding to the potential biological significance of this relationship is that both genes vary widely across the tumors in both datasets – BENE undergoes five doublings between the least and greatest expression, and Hs.143288 undergoes more than two doublings.

This pair of genes also illustrates the value of selecting genes based on differential correlation in addition to inspecting genes with differential mean expression. Neither BENE nor Hs. 143288 is significantly differentially expressed in mean between the two classes of samples, thus neither would be considered to be related to grade, based on usual measures of differential mean expression such as t-tests or fold-change statistics.

The BENE gene codes for a membrane-bound protein with unknown molecular function, and the Hs.143288 gene codes for a hypothetical protein with sequence similarity to mouse, rat, and *C.elegans* collagen. With little biological information, it is difficult to propose a mechanistic explanation for this relationship. One hypothetical explanation might be that advanced tumors segregate into two distinct clusters – one exhibiting high BENE expression and low Hs.143288 expression, and the other exhibiting high Hs.143288 expression and low BENE expression. This would suggest a permanent silencing of either BENE of Hs.143288 expression in all advanced tumors (but not a silencing of both genes in any one tumor). An alternative hypothetical explanation would be that both genes are transiently expressed in advanced tumor cells, but the expression is coordinated so that the two genes are never expressed simultaneously. This coordination may be associated with phenotypes such as proliferation, invasiveness, or vascularization that are more prominent in advanced tumors.

3.3.3 Genes participating in widespread differential correlation

Although many genes are differentially correlated with at least one other gene, we found that a few genes dominate all others, in that they participate in widespread differential correlation with many other genes. These genes may potentially play a more global role in reporting, or causing, widespread alterations in the interaction of gene expression levels. At a more practical level, they may serve as biomarkers for detecting dramatic shifts in correlation structure associated with a clinical endpoint.

For example, using tumor stage as the outcome, five genes engage in differential correlation with at least 20 other genes. Two of these genes are implicated in other epithelial adenocarcinomas, specifically, disease of ovary (WFDC2/HE4; Hs.2719) and colon (galectin-4; Hs.5302). Among the remaining three genes are a gene associated with female fertility (NRIP1; Hs.155017), a widely-expressed enzyme (MTHFD2; Hs.154672), and a gene of unknown function (Hs.380833).

Although galectin-4 is generally reported as being expressed only in colon, many of the lung tumors exhibit moderate expression of this transcript. While cross-hybridization of a different transcript is a likely explanation for this, given that we identified galectin-4 based on its differential correlation with respect to stage, it is notable that galectin-4 has been specifically noted as being associated with stage in colon cancer [Nagy et al., 2003].

The WFDC2/HE4 gene has been reported to be a biomarker for ovarian cancer [Hellstrom et al., 2003]. Expression in other tissues has been observed as well. Notably, high expression of WFDC2/HE4 is primarily found in malignant ovarian tumors, and it is expressed at much lower levels in non-malignant tumors. Since degree of malignancy is roughly associated with tumor stage, the specific expression of WFDC2/HE4 in malignant ovarian tumors may possibly be related to the fact that we found WFDC2/HE4 to be differentially correlated with stage in lung tumors.

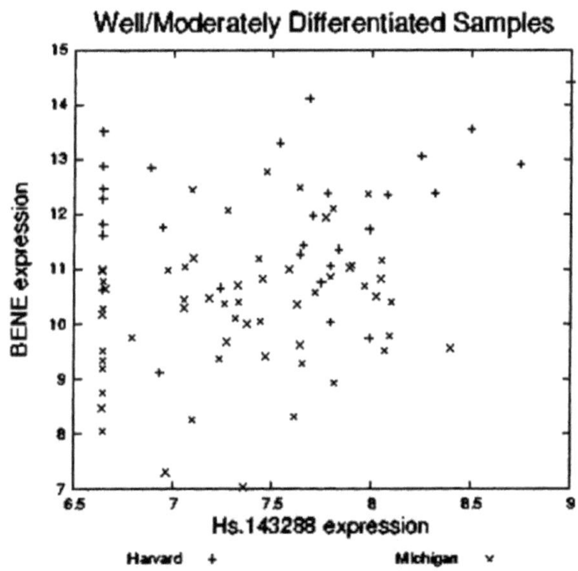

Figure 1. Moderately and well-differentiated samples show no association between Hs.143288 and BENE expression.

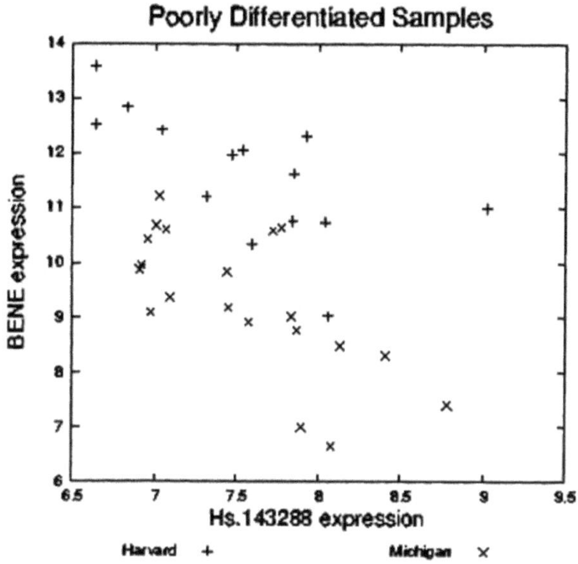

Figure 2. Poorly differentiated samples show negative association between Hs.143288 and BENE expression.

4. DISCUSSION

We have proposed a statistical analysis strategy for gene expression data that aims to uncover subtle associations between gene co-expression and important clinical outcomes. Our investigation of two lung-cancer data sets suggests that widespread statistically significant effects can be discovered based on differential correlation that would be missed using an analysis strategy focusing exclusively on differential mean expression. In particular, numerous genes may be associated with stage, grade, smoking status, and K-Ras mutation status that would not be detected using conventional measures of differential expression.

Differential correlation may be considered either in the context of a single data set, or, as we have done here, when looking at multiple data sets. In the latter case, we identified pairs of genes that are consistently differentially correlated in the two data sets. As with any statistical analysis, genes identified by differential correlation will be less likely to be attributable to experimental bias or noise if the results are reproduced in multiple data sets. In addition, differential correlation may be especially suitable for use with multiple data sets -- since differential correlation depends only on relationships between the expression levels of two genes, and not on absolute levels of gene expression, it is likely to be less sensitive to certain types of scaling artifacts that may produce systematic differences between results obtained in different laboratories or different microarray platforms.

We note that we have not had success in one of our primary goals, which was to enlarge the set of genes associated with survival using the differential correlation technique. Our baseline analysis suggests that only a small number of genes are associated with early death in both data sets, and none of these genes has a magnitude change greater than 1.5. No pair of genes shows significant differential correlation associated with early death.

One practical drawback of our method is that the response variable must be dichotomous, so that it can be used to stratify the samples into two classes. In some cases, such as when the response variable is survival time, this requires coarsening the resolution of the measurement, perhaps leading to a loss of relevant information. We note that a similar, but more general methodology called *Liquid Association* [Li, 2002] has recently been developed that has similar goals as our method, but is formulated so that a continuous rather than a binary outcome variable controls the changes in correlation.

5. ACKNOWLEDGMENTS

Our thanks to Rork Kuick for preparing the probeset-level data summaries, and for reviewing the manuscript.

6. REFERENCES

Beer DG, Kardia SL, Huang CC, Giordano TJ, Levin AM, Misek DE, Lin L, Chen G, Gharib TG, Thomas DG, Lizyness ML, Kuick R, Hayasaka S, Taylor JM, Iannettoni MD, Orringer MB, Hanash S. Gene-expression profiles predict survival of patients with lung adenocarcinoma. *Nature Medicine* 9 (816), 2002.

Bhattacharjee A, Richards WG, Staunton J, Li C, Monti C, Vasa P, Ladd C, Beheshti J, Bueno R, Gillette M, Loda M, Weber G, Mark EJ, Lander ES, Wong W, Johnson BE, Golub TR, Sugarbaker DJ, and Meyerson M. Classification of human lung carcinomas by mRNA expression profiling reveals distinct adenocarcinoma subclasses. *PNAS.* 98 (24), 13790-13795, November 2001.

Hellstrom I, Raycraft J, Hayden-Ledbetter M, Ledbetter JA, Schummer M, McIntosh M, Drescher C, Urban N, Hellstrom KE. The HE4 (WFDC2) protein is a biomarker for ovarian carcinoma. *Cancer Res.* 2003 Jul 1;63(13):3695-700.

Li KC Genome-wide coexpression dynamics: Theory and application. *PNAS* 2002 99 16875-16880.

Nagy N, Legendre H, Engels O, Andre S, Kaltner H, Wasano K, Zick Y, Pector JC, Decaestecker C, Gabius HJ, Salmon I, Kiss R. Refined prognostic evaluation in colon carcinoma using immunohistochemical galectin fingerprinting. *Cancer.* 2003 Apr 15;97(8):1849-58.

Wilcox R (1997). Introduction to Robust Estimation and Hypothesis Testing. Academic Press, New York.

This page intentionally left blank

Chapter 10

PROBABILISTIC LUNG CANCER MODELS CONDITIONED ON GENE EXPRESSION MICROARRAY DATA

Craig Friedman,[1] Wenbo Cao,[2] and Cheng Fan[3]
[1]*NYU Courant Institute of Mathematical Sciences;* [2]*City University of New York;* [3]*Lineberger Comprehensive Cancer Center, University of North Carolina at Chapel Hill*

Abstract: A number of quantitative methods have been applied to the classification and clustering of microarray data (see, for example, [Tibshirani et al., 2001]). In this article, we describe a statistical learning theory-based method to construct lung cancer probability models that are conditioned on gene expression microarray data. Our models do more than classify—they indicate an estimate of the probability. We find our estimate for the conditional probability distribution by choosing a model that balances consistency with the training data and consistency with a prior distribution. This formulation leads to an optimization problem that has a mathematically equivalent problem with an objective function that is a penalized log-likelihood. We discuss three particular estimation problems: 1) find the conditional probability that a sample is adenocarcinoma or normal, given gene expression levels, 2) find the conditional probability for each of six disjoint categories related to lung cancer, given gene expression levels, and 3) find the conditional probability distribution for survival time, given gene expression levels. We describe the features that we select and measure the performance of the models that we create in economic terms. For the conditional probability of adenocarcinoma, we condition on probeset identifiers common to both the Harvard and Michigan data sets. When we trained on either data set, we were able to nearly perfectly classify adenocarcinoma on the other set.

Key words: Microarray, ontology, adenocarcinoma, conditional probability, gene expression, features

1. INTRODUCTION

DNA microarrays can be used to characterize the molecular variations among tumors by monitoring gene expression profiles on a genomic scale. Supervised learning (including classification, discriminant analysis, and supervised pattern recognition) of DNA microarray gene expression data may prove useful in cancer research for the purpose of tumor classification. Popular supervised learning methods include linear discriminant analysis, k-nearest neighbor classifiers, nearest centroid classification, classification trees, and support vector machines [Dudoit et al., 2002]. A set of labeled objects (training set: a group of patients with known gene expression profiles and clinical outcomes) is used to build a learning machine that can predict the class of future unlabeled objects (testing set: patients with known gene expression profiles, but without clinical outcomes).

In this paper, we use a utility-based approach to estimate the conditional probability of human lung cancer, given DNA microarray gene expression levels. Our method incorporates automated feature selection. The models produced by this method provide more information than mere classification—they explicitly produce probabilities estimates, which quantify our confidence in the classification. The threshold for the classification is flexible; this may lead to a more reliable and precise identification of tumors and provide more specificity for diagnosis and treatment, like personalized medicines.

We have used two of the four CAMDA03 contest data sets for our study: the CAMDA03 Harvard data set [Bhattacharjee et al., 2001], and the CAMDA03 Michigan data set [Beer et al., 2002]. The Harvard data set was measured by the Hu_U95Av2 Affymetrix chip, while the Michigan data set was measured by the HG_GeneFL Affymetrix chip. The Hu_U95Av2 chip contains 12,625 probesets, each with 16 probe pairs. The HG_GeneFL chip is older than the Hu_U95Av2 chip, and only contains 6,633 probesets, each with 20 pairs.

2. PROBLEMS AND SOLUTION METHODOLOGY

In this section, we describe the probabilistic models that we seek, our solution methodology, and the features that we use for three specific models: 1) a conditional probabilistic model for adenocarcinoma or normal tissue, given gene expression level data, 2) a conditional probabilistic multicategory model for six disjoint categories of tissue (normal lung, adenocarcinoma, small-cell lung carcinomas, squamous cell lung carcinomas, pulmonary carcinoids, and other adenocarcinomas--which were suspected to be extrapulmonary metastases) and 3) a conditional survival time model, given gene expression level data.

Let x denote our vector of explanatory variables and the random variable Y denote the state of the random variable for which we seek the conditional (on x) probability. We seek models for the probability distribution of Y, given x.

For our first numerical experiment, $Y \in \{0,1\}$ indicates adenocarcinoma (Y=l) or normal lung (Y=0). We seek the conditional probability measure $p(1|x)=Prob(Y=1|x)$.

For our second numerical experiment, there are six states: $Y = \{(0,0,0,0,0,1),(0,0,0,0,1,0),..., (1,0,0,0,0,0)\}$, which represent normal lung, adenocarcinoma, small-cell lung carcinomas, squamous cell lung carcinomas, pulmonary carcinoids, and other adenocarcinomas (which were suspected to be extrapulmonary metastases), respectively. We seek the conditional probability measure $p(y|x)=Prob(Y=y|x)$, where $y \in Y$.

For our third numerical experiment, $Y \in (0,\infty)$ indicates survival time. In this case, we seek the conditional probability density $p(y|x)$.

In each case, we use the maximum expected utility (MEU) methodology (described in the Appendix) to estimate $p(y|x)$. This method requires that we specify features, which are functions of x and y that can be thought of in two ways described more precisely in the Appendix: 1) they are akin to a basis set of functions that we can search over and combine to form models, and 2) they are used to enforce consistency of the model with the data--the introduction of more and more features allows for more and more feature constraints, which induces more and more consistency of the model with the data. In this article, we use three particular types of features: linear features, quadratic features, and Gaussian kernel features, which are described precisely in the Appendix. Linear and quadratic features are akin to the first two term types in a multidimensional Taylor expansion. The Gaussian kernel features allow for local behavior and depend on a bandwidth hyperparameter. Another hyperparameter, α, also defined in the appendix, is used to mitigate overfitting.

3. EXPLANATORY VARIABLES AND FEATURES

For the conditional adenocarcinoma (or normal) model, we use three types of features: linear features, quadratic features, and Gaussian kernel features. In order to integrate information between different versions of chips, while performing out-of-sample tests in a natural way, we trained two models. One model was trained on the Harvard data set and tested on the Michigan data set (the Harvard-Michigan experiment); the other was trained on the Michigan data set and tested on the Harvard data set (the Michigan-Harvard experiment). The Harvard data set contained data for 127 lung adenocarcinoma samples, 12 suspected extrapulmonary metastases samples, and seven normal lung samples. The Michigan data set contained data for 86 lung adenocarcinoma samples and 10 normal lung samples. In order to find the correspondence between the two data sets, we first mapped probesets to UNIGENE ID's by using the current annotation files for HG_GeneFL and Hu_U95Av2 chips. Then, we grouped probesets with the same UNIGENE ID together. We found 5,001 groups for the Michigan data set, and 8,778 groups for the Harvard data set. In the third step, we calculated expression levels for each group by averaging the expression levels of all probesets belonging to the same group. Finally, by using UNIGENE ID's, we found groups common to both the Michigan and Harvard data sets. Ultimately, we obtained 4,822 groups common to both the Michigan and Harvard data sets. In order to compensate for the measurement differences associated with the different chips, we rank transformed the expression levels.

```
X0 ← ∅
X1 ← {x1, x2,..., xd}
for m ← from 1 to n do
   Select one component from X1 which can be used with
    components in X0 to achieve highest log-likelihood, say xk
   X0 ← X0 ∪ {xk}
   X1 ← X1 \ {xk}
end
return X0
```

Figure 1. Algorithm for selecting significant genes.

In order to choose prognostic genes, we followed the following procedure: in the first step, we chose the explanatory variable with the greatest log-likelihood. In the second step, we choose another one which can be used with the already chosen ones to achieve the highest log-likelihood. We repeated this procedure until a certain number of components of X (the set of explanatory variables) had been selected. Pseudo-code for this algorithm is given in Figure 1. In Figure 1, $\mathbf{X}_0$ is the set of selected components, $\mathbf{X}_1$ is the set of all candidate components to be selected, and n is the number of components we want to select. Initially $\mathbf{X}_0$ was an empty set and $\mathbf{X}_1$ contained all of the components of X. For computational efficiency, we used linear features only in the selection procedure. Using this algorithm, we selected the 10 and eight most significant genes for Harvard-Michigan experiment and Michigan-Harvard experiment, respectively (see Table 1 and Table 2). Three genes, TNA, POLR2H, and CMRF35 appear in the selected gene lists for both experiments. Consequently, we have 76 features for Harvard-Michigan experiment, of which 11 are linear features, 55 are quadratic features, and 10 are Gaussian features; we have 55 features for Michigan experiment, of which nine are linear features, 36 are quadratic features, and 10 are Gaussian features.

Table 1. Selected genes in Harvard-Michigan experiment.

Rank	*Gene Symbol*	*Gene Name*
1	*TNA*	*tetranectin (plasminogen binding protein)*
2	*POLR2H*	*polymerase (RNA) II (DNA directed) polypeptide H*
3	*CMRF35*	*CMRF35 leukocyte immunoglobulin-like receptor*
4	*F8*	*coagulation factor VIII, procoagulant component (hemophilia A)*
5	*HLCS*	*holocarboxylase synthetase (biotin-[proprionyl-Coenzyme A-carboxylase (ATP-hydrolysing)] ligase)*
6	*HLA-DRB3*	*major histocompatibility complex, class II, DR beta 3*
7	*MEST*	*mesoderm specific transcript homolog (mouse)*
8	*HSPA1B*	*heat shock 70kDa protein 1B*
9	*CEBPG*	*CCAAT/enhancer binding protein (C/EBP), gamma*
10	*LILRB4*	*leukocyte immunoglobulin-like receptor, subfamily B (with TM and ITIM domains), member 4*

Table 2. Selected genes in Michigan-Harvard experiment.

Rank	*Gene Symbol*	*Gene Name*
1	*TNA*	*tetranectin (plasminogen binding protein)*
2	*POLR2H*	*polymerase (RNA) II (DNA directed) polypeptide H*
3	*CMRF35*	*CMRF35 leukocyte immunoglobulin-like receptor*
4	*TTC1*	*tetratricopeptide repeat domain 1*
5	*CDH18*	*cadherin 18, type 2*
6	*MGP*	*matrix Gla protein*
7	*ROS1*	*v-ros UR2 sarcoma virus oncogene homolog 1 (avian)*
8	*RBM5*	*RNA binding motif protein 5*

For the conditional six state multicategory model, we use three types of features: linear features, quadratic features, and Gaussian kernel features. The Harvard data set contains six different types of data. Three were mentioned above; the other three types are: squamous cell lung carcinomas (21 samples), pulmonary carcinoids (20 samples), and small-cell lung carcinomas (six samples). We built a six state conditional probability model that generates probabilities for each of the six different types of data, given gene expression levels.

Selecting prognostic probesets from 12,600 candidates is very time-consuming. To reduce the time complexity of the selection procedure, we first divide 12,600 probesets into 630 groups, each with 20 probesets; from each group, the two most significant probesets are selected by the algorithm described above. In the second step, we divided the 1,260 selected probesets from the first step into 63 groups, each with 20 probesets; again, we selected the two most significant probesets from each group. In the third step, we chose the 10 most significant probesets from 126 chosen ones from the second step. Of these 10 chosen probesets, we removed two that correspond to non-human genes and one that corresponds to a housekeeping gene. (It is possible that such bacterial gene expression data may reflect different handling of tumor and normal tissues; this possibility was suggested to us by Michael Ochs and attributed to Giovanni Parmigiani.) Finally we selected seven probesets (see Table 3) as input for our conditional six state multicategory model. From these seven probesets, we generated 276 features: 48 linear features, 168 quadratic features, and 60 Gaussian features.

Table 3. Selected probesets and their UniGene ID's and LocusLink's

Rank	*Gene Symbol*	*Gene Name*
1	*KCNK3*	*potassium channel, subfamily K, member 3*
2	*RFC4*	*replication factor C (activator 1) 4, 37kDa*
3	*DSP*	*desmoplakin*
4	*GPX3*	*glutathione peroxidase 3 (plasma)*
5	*INSM1*	*insulinoma-associated 1*
6	*FCGRT*	*Fc fragment of IgG, receptor, transporter, alpha*
7	*TNKS*	*tankyrase, TRF1-interacting ankyrin-related ADP-ribose polymerase*

For the conditional survival time model, using the clinical history information from the Michigan data set, we performed an experiment to predict the conditional probability density for survival time for each patient. We selected ten probesets from 7,129 by a procedure similar to that described in above. After removing two probesets that corresponded to non-human genes, we had eight probesets (see Table 4). Since only 79 samples were available, we used only linear features in our numerical experiment. We used only nine linear features.

Table 4. Selected probesets, and their UniGene ID's and LocusLink's

Rank	*Gene Symbol*	*Gene Name*
1	*TRA2A*	*transformer-2 alpha*
2	*PPP5C*	*protein phosphatase 5, catalytic subunit*
3	*MATN1*	*matrilin 1, cartilage matrix protein*
4	*CA4*	*carbonic anhydrase IV*
5	*ZNF3*	*zinc finger protein 3 (A8-51)*
6	*RPL27A*	*ribosomal protein L27a*
7	---	*Homo sapiens cDNA FLJ30561 fis, clone BRAWH2004580.*
8	*UROD*	*uroporphyrinogen decarboxylase*

4. MODEL PERFORMANCE MEASUREMENT AND RESULTS

To measure the performance of our model, p, we use Δ, the scaled log-likelihood difference between our model and a benchmark model as estimated on an out-of-sample dataset

For the conditional adenocarcinoma (or normal) model, to compute our performance measure, we train on one of the data sets (Harvard or Michigan) and calculate the performance measure on the other. For each model, we also display the area under the receiver operator characteristic (ROC) curve a popular, rank-based performance measure. We take as a benchmark model the linear logistic regression. We note that linear logistic regression is a special case of our approach when the prior is flat, $\alpha=0$, and the features are linear. Thus, our approach can be viewed as a generalization which is better able to handle nonlinearities and overfitting. It is easy to show that, for the linear logit model, the level sets of the conditional probability of adenocarcinoma surface must satisfy a rather strict geometrical condition: they must be linear. This imposes a severe restriction on the model, which may not be sufficiently flexible to conform to the story told by the data.

Our model depends on hyperparameters (the regularization factor α and bandwidth σ) and parameters (the β vector). We test a discrete set of (α,σ) pairs. For each pair, we trained our model on either the Harvard data set or the Michigan data set, and calculated Δ on the same data set. We selected hyperparameter values corresponding to the greatest entry in the Δ table. We then used the resulting model to evaluate performance measures on the other data set. We display model performance statistics (against the noninformative model) in Table 5 and model-produced probabilities for the Harvard-Michigan experiment in Figure 2. The MEU method produces high

ROC values on either data set. We note that though MEU has good performance on both data sets, the logistic regression has better performance on Harvard-Michigan, perhaps due to the fact that the populations are, in fact, not the same and logistic regression, though not as precisely tuned to the Harvard set, performs better on the Michigan set. The difference might reflect the need for a platform normalization to adjust for the batch bias.

Table 5. Model performance statistics for MEU V.S. logistic regression.

Experiment	*Model Measurement*	*MEU*	*Logistic Regression*
Harvard-Michigan	Δ	*0.1552*	*0.3335*
	ROC	*0.9837*	*1.0*
Michigan-Harvard	Δ	*0.2719*	*-17.8833*
	ROC	*0.9928*	*0.8324*

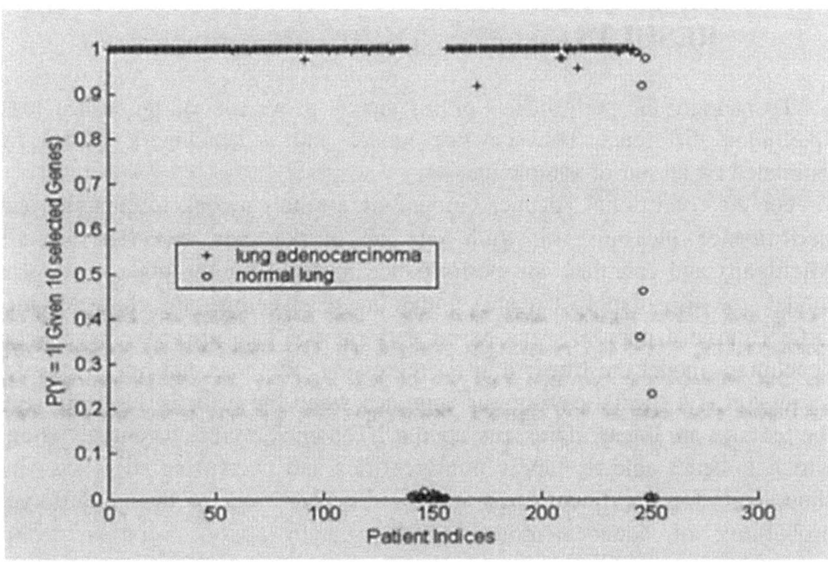

Figure 2. Conditional probability of adenocarcinoma given 10 selected genes (Harvard-Michigan experiment).

According to Hosmer and Lemeshow [2000], the area under the *ROC* curve can be interpreted as the percentage of (AD,NL) outcome pairs that are ranked correctly by the model, and, "As a general rule: If ROC = 0.5: this suggests no discrimination.... If $ROC \geq 0.9$: this is considered outstanding discrimination. In practice it is extremely unusual to observe areas under the ROC curve greater than 0.9."

We have produced two models, one trained on the Harvard data, the other trained on the Michigan data. In either case, the MEU model produced nearly perfect classification on the out of sample data sets, as is evident from Table 5 and Figure 2.

For the conditional six state multicategory model, we randomly designated 80% of the Harvard data as training data and the remaining 20% as test data. For each given regularization factor, α, and bandwidth, σ, we trained our model on the training data and calculated Δ on the test data. We selected hyperparameter values corresponding to the greatest entry in the Δ table. We then used the resulting model to evaluate performance measures on the test data set. We repeated this procedure 30 times and reported Δ as the mean value of all 30 Δ's. This model was benchmarked against the noninformative model and produced $\Delta = 0.8012$ ($\Delta = -2.2166$ for using linear features only, without regularization). We display model probabilities in Figures 3 and 4.

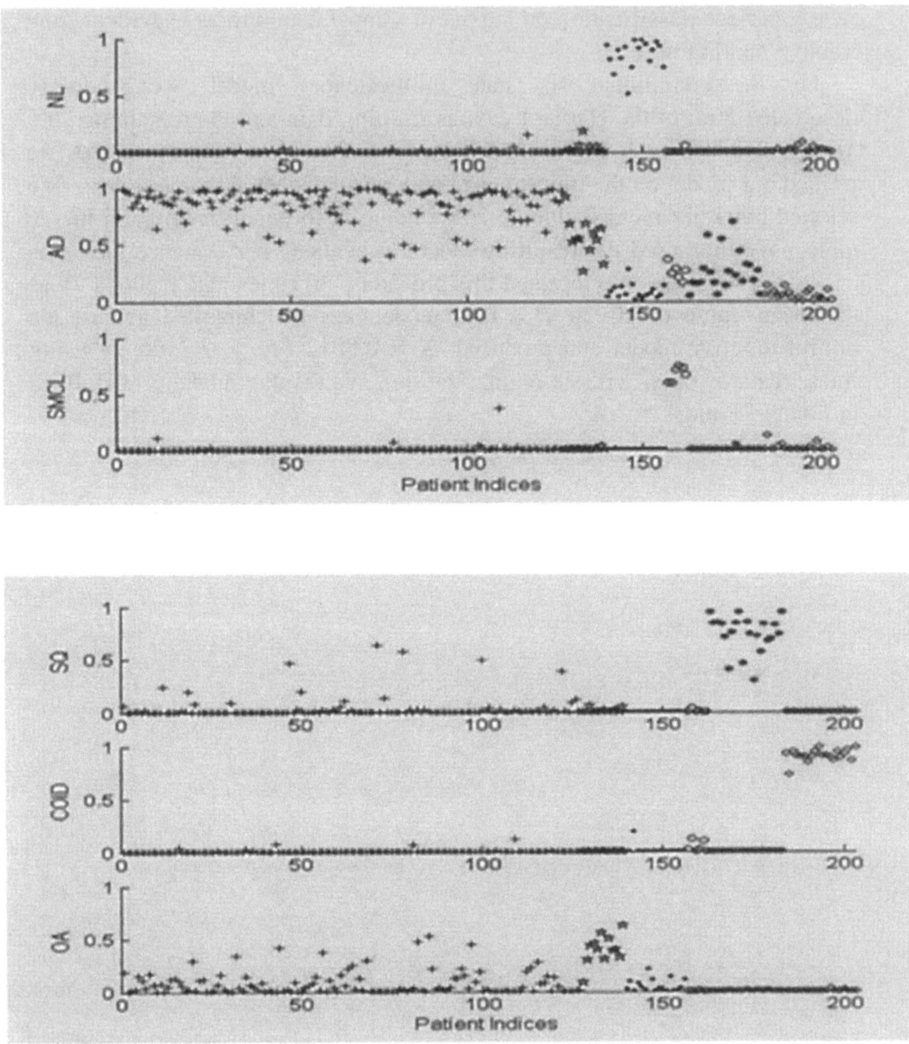

Figure 3. Conditional six state multicategory experiment (1Legend description: normal lung (NL): dot; lung adenocarcinomas (AD): plus; small-cell lung carcinomas (SMCL): circle; squamous cell lung carcinomas (SQ): asterisk; pulmonary carcinoids (COID): diamond; other adenocarcinomas (OA): pentagram).

For the conditional survival time model, survival time, measured in months, is the patient's survival time from the operation date to death or last follow up as of May, 2001. We randomly designated 80% of the Michigan data as training data and the remaining 20% as test data. We then trained our model, without regularization, on the training set and calculated the performance measure Δ on the test data. We repeated this procedure 30 times, and reported $\Delta = 0.3057$ as the mean value of all calculated Δ's. We display a few conditional survival time probability distributions in Figure 4. Without regularization, our model is equivalent to a maximum likelihood exponential model and we handle this censored type I data via maximum likelihood estimation (see NIST/SEMATECH [2004]). It was not necessary to regularize since there were sufficiently many data, given the number of features. With more features, regularization might have been beneficial.

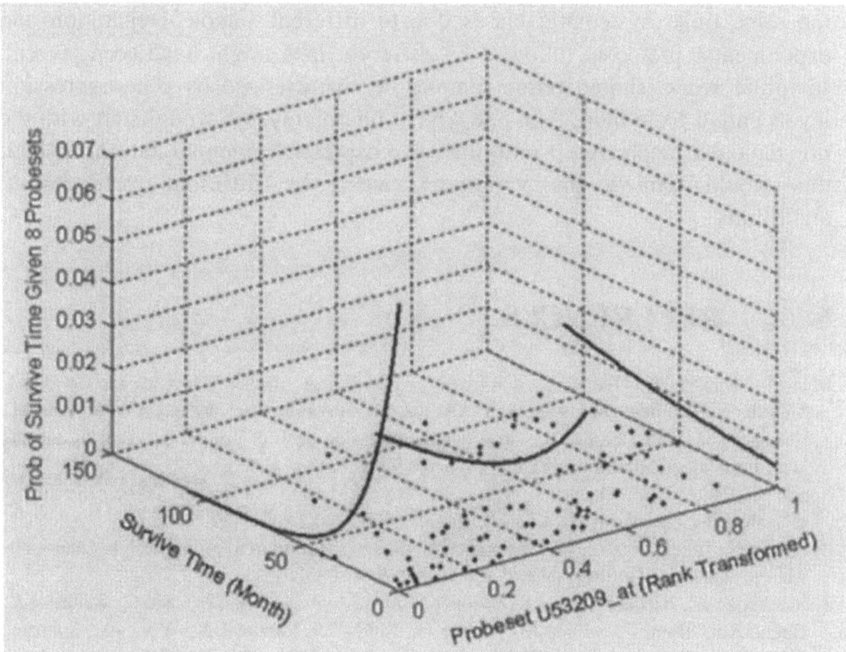

Figure 4. Conditional PDF as a function of selected probesets and data. (values for probesets other than U53209_at are taken as median of their rank-transformed values).

5. DISCUSSION

In the course of our automated training on the Harvard data set and testing on the Michigan data set, 10 genes with predictive power were selected. In the course of our automated training on the Michigan data set and testing on Harvard data set, eight genes with predictive power were selected. There are three common genes.

Table 1 and Table 2 list these genes; a number of them have received attention and have been previously reported in human lung carcinoma research [Bhattacharjee et al., 2001;Berglund and Linell, 1972; Gure et al., 1998]. We also observed several previously uncharacterized biomarkers for lung cancer, which we believe deserve further study and validation.

In order to measure gene expression, we must preprocess the probe level intensity data from high-density oligonucleotide arrays. Different expression summarization methods yield different results. We have used the gene expression data sets preprocessed by Harvard and Michigan, respectively. We do not know if they used the same expression summarization method. At the same time, systematic biases due to different sample preparation and experimental protocols followed by different labs might have been present. In future work, should tissue samples be characterized by gene expression levels culled from more than one type of microarray, we would start with the raw data and preprocess it with the same expression summarization method; this would remove the variations caused by different preprocessing algorithms.

6. REFERENCES

Beer, D., Kardia, S., Huang, C., Giordano, T., Levin, A., Misek, D., Lin, L., Chen, G;, Gharib, T., Thomas, D., Lizyness, M., Kuick, R., Hayasaka, S., Taylor, J;, Iannettoni, M., Orringer, M., and Hanash, S., 2002, Gene-expression profiles predict survival of patients with lung adenocarcinoma, *Nature Medicine* **9**:816.

Benito M., Parker J., Du Q., Wu J., Xiang D., Perou C.M., and Marron J.S., 2004, Adjustment of systematic microarray data biases, *Bioinformatics* **20**(1): 105-114.

Berglund, S.; Linell, F.1972. Fibrosis and carcinoma of the lung in a family with haemoglobin Malmo--anatomic findings, *Scand. J. Haemat.* **9**: 424-432.

Bhattacharjee A., Richards W.G., Staunton J., Li C., Monti S., Vasa P., Ladd C., Beheshti J., Bueno R., Gillette M., Loda M., Weber G., Mark E.J., Lander E.S., Wong W., Johnson B.E., Golub T.R., Sugarbaker D.J., and Meyerson M., 2001, Classification of human lung carcinomas by mRNA expression profiling reveals distinct adenocarcinoma subclasses, *Proc Natl Acad Sci U S A.* **98**(24):13790-5.

Dudoit, S., Fridlyand J., and Speed T.P., 2002, Comparison of discrimination methods for the classification of tumors using gene expression data, *J. Am. Statistical Assoc.,* **97**(457):77-87.

Friedman, C., and Sandow, S., 2003a, Learning probabilistic models: an expected utility approach, *Journal of Machine Learning Research,* **4**:257-291.

Friedman, C., and Sandow, S., 2003b, Ultimate recoveries, *Risk,* **16**(8):69-73.

Gure, A. O., Altorki, N. K., Stockert, E., Scanlan, M. J., Old, L. J., Chen, Y.-T., 1998, Human lung cancer antigens recognized by autologous antibodies: definition of a novel cDNA derived from the tumor suppressor gene locus on chromosome 3p21.3. *Cancer Res.* **58**: 1034-1041.

Hosmer, D., and Lemeshow, S., 2000, *Applied Logistic Regression,* Second edition Wiley, New York.

Jaynes, E., 1957, Information theory and statistical mechanics, *Physical Review,* **106**:620.
Lebanon, G., and Lafferly, J., 2001, Boosting and maximum likelihood for exponential models, in *Advances in Neural Information Processing Systems,* **14,** MIT Press, Cambridge, Ma.
Lin, S., and Johnson, K., 2002, *Methods of Microarray Data Analysis II,* Kluwer, 2002.
NIST/SEMATECH, 2004 *e-Handbook of Statistical Methods,* http://www.itl.nist.gov/div898/handbook/
Tibshirani, R., Hastie, T., Narasimhan, B., and Chu, G., 2002, Diagnosis of multiple cancer types by shrunken centroids of gene expression, PNAS **99**(10): **6567-6572**

APPENDIX: MEU METHODOLOGY

We follow the maximum expected utility (MEU) modeling approach from [Friedman and Sandow, 2003a]. This methodology is designed to take into account the decision consequences for a decision maker who relies on the model to make decisions, with a hyperparameter, α, governing the balance between consistency with prior beliefs (the model that we believe *before* we observe data) and consistency with the data. This methodology admits models that are flexible enough to conform to the data, yet avoid overfitting. Under this methodology, we estimate $p(y|x)$ by maximizing, over $\beta=(\beta_1,\ldots,\beta_J)^T$, the regularized maximum likelihood

$$h(\beta)=\frac{1}{N}\sum_{k=1}^{N}\log p^{\beta}(y_k\mid x_k)-\frac{\alpha}{N}\beta^T\Sigma\beta \tag{1}$$

with

$$p^{\beta}\left(y\mid x\right)=\frac{1}{Z_x(\beta)}e^{\beta^T f(y,x)}p^0\left(y\mid x\right) \text{ and} \tag{2}$$

$$Z_x(\beta)=\sum_y p^0\left(y\mid x\right)e^{\beta^T f(y,x)} \tag{3}$$

(if there are an infinite number of Y states, as for the survival time problem, we can replace the sums in Eq. (7) by an integral—see, for example, Friedman and Sandow [2003b]) where the (x_k, y_k) are the observed (x,y)-

pairs, N is the number of observations, $\mathbf{f}(y,x)=(f_1(y,x),\ldots,f_J(y,x))^T$ is the vector of features and $\mathbf{\Sigma}$ is the empirical covariance matrix of the features. The features are functions of x and y that can be thought of in two ways: 1) from Eqs. (2) and (3), we see that our solution can be expressed in terms of $\beta-$ weighted linear sums of features, so the set of features is akin to a basis set of functions that we can search over to form models, and 2) they are used to enforce consistency of the model with the data, since model expected feature values are forced to be approximately equal to empirical feature expectations (see Friedman and Sandow [2003a], Problem 1)--the introduction of more and more features allows for more and more feature constraints, which induces more and more consistency of the model with the data. In this article, we use at most three particular types of features: linear features $f_j(y,x)=(y-c)(x)_j$ (where $(x)_j$ denotes the jth coordinate of x, with the convention that $(x)_0=1$, and c=0 for the survival time problem and .5 otherwise), quadratic features $f_j(y,x)=(y-.5)(x)_i(x)_j$, and Gaussian kernel features $f_j(y,x)=(y-.5)\exp(-\sigma\|x-x^k\|^2)$ (where x^k is one of K centers selected via the method of k-mean centers and σ is a bandwidth hyperparameter). Linear and quadratic features are akin to the first two types of terms in a Taylor expansion. The Gaussian kernel features allow for local behavior.

In our numerical experiments, the results were somewhat insensitive to the prior measure; the results described were obtained for the noninformative prior measure (the observed *unconditional* probability that $Y = y$).

Chapter 11

INTEGRATION OF MICROARRAY DATA FOR A COMPARATIVE STUDY OF CLASSIFIERS AND IDENTIFICATION OF MARKER GENES

Daniel Berrar, Brian Sturgeon, Ian Bradbury, C. Stephen Downes, and Werner Dubitzky
School of Biomedical Sciences, University of Ulster at Coleraine, Northern Ireland

Abstract: Novel diagnostic tools promise the development of patient-tailored cancer treatment. However, one major step towards individualized therapy is to use a combination of various data sources, e.g. transcriptomic, proteomic, and clinical data. We have integrated clinical data and lung cancer microarray data that were generated on two different oligonucleotide platforms. We were interested in the question whether the prediction of survival outcome benefits from the integration of clinical and transcriptomic data. In addition, we attempted to identify those genes whose expression profiles correlate with survival outcome. We applied five machine learning techniques to predict survival risk groups, and we compared the models with respect to their performance and general user acceptance. Based on quantitative and qualitative evaluation criteria, we chose decision trees as the most relevant technique for this type of analysis. Our *in silico* analysis corroborates the role of numerous marker genes already described in lung adenocarcinomas. In addition, our study reveals a set of highly interesting genes whose expression profiles correlate with genetic risk groups of unexpected survival outcomes.

Key words: Microarray, lung cancer, survival analysis, machine learning

1. INTRODUCTION

Modern high-throughput technologies produce growing amounts of biomedical data. Transcriptional profiling using microarray technology promises to uncover unprecedented insights into the pathogenesis of complex diseases such as cancer. Recent studies on cancer profiling have demonstrated that gene expression patterns of cancer can be successfully used for survival prognosis, e.g., in childhood leukemia [Yeoh et al., 2002], lung cancer [Bhattacharjee et al., 2001; Beer et al., 2002], and breast cancer [van't Veer et al., 2002]. Delineating cancers based on their specific expression profiles may provide the breakthrough required to develop a patient-tailored therapy. Currently, it is unclear how individual patients respond to chemotherapy. Existing chemotherapies have in general severe side effects for the patients, but sometimes low efficacy.

Supervised machine learning techniques are a promising approach for analyzing microarray data in the context of patient outcome prediction. For example, Shipp et al. [2002] reported on the successful survival prediction of patients suffering from large B-cell lymphoma. They employed machine learning techniques (support vector machine, *k*-nearest neighbor) to predict the survival periods of a group of patients. It was shown that the predictive accuracy based on the expression profiles was higher than that based on simple clinical parameters.

Despite the undisputable credentials of microarray technology, transcriptional profiling alone is insufficient to explain the whole spectrum of alterations involved in cancer genesis. Combining gene expression data with proteomic data, cytogenetic data (e.g., from fluorescence in situ hybridization experiments), and clinical patient data might be a promising approach for developing new prognostic tools [Ochs et al., 2003]. Particularly in the context of cancer outcome prediction, the integration of heterogeneous data sources is considered to be a promising new approach. The question whether decision support systems based on machine learning approaches and microarray data will find their way into clinical practice is still open, and many other problems remain unresolved. One of the main bioinformatics challenges is the integration of heterogeneous data sources and the development of methods and tools for analyzing high-dimensional microarray data.

2. STUDY OUTLINE

In the present study, we have analyzed the Harvard lung cancer data set [Bhattacharjee et al., 2001] and the Michigan lung cancer data set [Beer et al., 2002]. Both data sets were generated on Affymetrix platforms. The data sets comprise a different number of genes and clinical parameters for the patients, but there exists a subset of genes that is contained in both data sets. The Harvard data set comprises expression data from 12,600 transcript sequences for 186 patients, including 139 adenocarcinomas. The Michigan data set contains 86 primary lung adenocarcinomas as well as 10 nonneoplastic lung samples. Furthermore, various clinical data are provided, such as tumor stage and anamnestic data (e.g., smoking habits, sex, age).

In the first part of our study, the classification task, we are interested in the question whether the combination of both data sets in conjunction with the clinical data can improve the prediction of the 5-year survival chance of patients. We decided to consider the 5-year survival prediction task because this question is clinically motivated and of practical relevance. The question is: Can we predict the survival risk group of the patients (survival < 5 years or survival ≥ 5 years)? Thus, we are essentially formulating the problem as two-class classification task. To address this task, we compare five state-of-the-art machine learning techniques.

In the second part, we are concerned with a regression task. We use a survival tree as an exploratory tool. Here, we are interested in the question of whether we are able to discover novel, non-trivial, and potentially useful (with clinical impact) insights into the correlation between gene expression and survival outcome.

3. DATA INTEGRATION

We developed a relational database to facilitate the integration and preprocessing of the heterogeneous data. The gene expression data for both sets was integrated using the common attribute or key *gene name.* In total, 3,588 genes are in common in both the Harvard and the Michigan data set. We selected only these genes for further analysis. Furthermore, we selected the following clinical and anamnestic parameters: age, sex, TNM classification, tumor stage, survival time in months, and censor index. For some patients in the Harvard data set no survival information was given. We excluded these patients from further analysis, so that the data set for analysis contained a total of 211 patients (125 from Harvard data set, 86 from Michigan data set). For the classification task, we excluded all patients that were censored before 5 years (75 patients). In the next step, we discretized the continuous values of the patients' survival data into two classes, high risk and low risk. Patients in the group high risk died before the 5 year mark, whereas patients in the group low risk survived at least 5 years after diagnosis. For the regression task, we included all 211 patients, i.e. both censored and uncensored observations.

4. DATA NORMALIZATION

Although both Harvard and Michigan data have been generated on an Affymetrix platform, the measured gene expression values are not directly comparable. The arrays used in the two studies have different probe sets, making a direct comparison of transcript abundance problematic. Beer et al. [2002] used HuGeneFL chips, containing 6,633 probesets, each with 20 probe pairs. Bhattacharje et al. [2001] used HG_U95Av2 chips, with 12,625 probesets and 16 probe pairs each. In addition, the survival outcome in the two data sets is significantly different ($p = 0.0049$, comparison of Kaplan-Meier curves). The Harvard data set contains 15 patients of tumor stage IV, while the Michigan data set contains no patients with metastasis.

A global normalization approach using a simple mean or median centering method is probably not sufficient because of the significantly different survival outcomes. It is therefore crucial to make the data of the two sets as comparable as possible by removing any set-dependent bias; otherwise, we cannot exclude the possibility that the marker genes are discriminating with respect to the data sets and not with respect to the survival outcome.

Bolstad et al. [2002] compared different normalization methods for oligonucleotide arrays and identified quantile normalization as a method of choice, both with respect to speed and variance and bias considerations. Using this method, it is possible to make the distribution of probe intensities for each array in a set of arrays approximately the same. We therefore adopted a quantile normalization method for integrating the two data sets. This normalization scheme consists of four steps. Let $\mathbf{v}_i$ and $\mathbf{v}_j$ be column vectors of expression values generated on two different oligonucleotide platforms.

(1) Sort $\mathbf{v}_i$ and $\mathbf{v}_j$ in ascending order and determine the quantiles.
(2) Using linear regression, impute the values for the quantiles that are in $\mathbf{v}_i$ but not in $\mathbf{v}_j$, and vice versa.
(3) For all k quantiles of expression values, compute the mean, m_k, of values in $\mathbf{v}_i$ and $\mathbf{v}_j$ in the kth quantile, and assign m_k to the kth element in $\mathbf{v}_i$ and $\mathbf{v}_j$.
(4) Rearrange $\mathbf{v}_i$ and $\mathbf{v}_j$ to the original order.

The vectors $\mathbf{v}_i$ and $\mathbf{v}_j$ are quantile-normalized expression vectors. The data were further normalized by standardizing each expression profile to mean 0 and variance 1. We performed a principal component analysis on the quantile-normalized expression data. The scatter plot of the first three principal components showed that the data cover approximately the same space.

5. DATA ANALYSIS

5.1 Material

To address the classification task, we split the data set of 136 patients randomly into a learning set of 96 patients (70.0%) and a test set of 40 patients (30.0%). The learning set comprises 63 (65.6%) patients of class high risk and 33 (34.4%) patients of class low risk. The test set comprises 26 (65.0%) patients of class high risk and 14 (35.0%) patients of class low risk. We include different variables in the data sets:

(1) *Patient Data:* age, sex, TNM-status, and tumor stage.
(2) *Expression Data:* quantile-normalized expression values.
(3) *Patient+Expression Data:* both *Patient Data* and *Expression Data.*

We trained the classifiers on the learning set and applied them to the test set with the corresponding set of variables.

5.2 Methods

In the classification task, we investigated the performance of five state-of-the-art machine learning methods: decision trees (C5.0), support vector machines (SVMs), probabilistic neural networks (PNNs), k-nearest neighbor classifier (k-NN), and artificial neural networks (multilayer perceptrons, MLPs). Recent studies have reported successful application of these methods to classification of microarray data, e.g., decision trees [Zhang et al., 2001], SVMs [Brown et al., 2000], PNNs [Berrar et al., 2003], k-NN [Shipp et al., 2002], and MLPs [Khan et al., 2001].

We assessed the models' performance on the basis of two criteria. Firstly, based on a quantitative criterion, the classification accuracy with respect to the survival risk groups. Secondly, on the basis of a qualitative criterion, the output interpretability, i.e. we are interested in the question: "How intelligible is the model's output to humans?" In the following, we briefly describe the classifiers.

SVMs belong to the family of binary statistical classifiers [Burges, 1998]. The basic principle of a SVM consists of finding the optimal separating hyperplane between two distinct classes.

MLPs belong to the family of artificial neural networks. In the present study, we use SPSS Clementine's implementation of MLPs and train the networks using the backpropagation algorithm.

A PNN is the parallel implementation of the Bayes-Parzen classifier [Specht, 1990].

Like the PNN, the *k*-NN classifier needs to access all learning data at the time when a new test case is to be classified. In its simplest implementation, the *k*-NN classifier computes a measure of similarity between the new case and all learning cases, and the new case is classified as a member of the same class as the most similar case. In the present study, we implemented a weighted *k*-NN that takes into account the similarity of the nearest neighbors for classifying a new case.

Based on some measure of "purity" such as information gain, decision trees recursively split the data set by selecting the features (genes) that are most discriminating with respect to the classes [Zhang et al., 2001]. For the present study, we applied SPSS Clementine's implementation of the decision tree C5.0. Survival trees belong to the family of classification and regression trees (CART), which operate similarly to the decision tree algorithm. The CART methodology builds a binary decision tree by recursively partitioning the elements of a data set according to some splitting rule. For an overview of classification and regression trees, see e.g. Breiman et al. [1984]. The splitting criterion for C5.0 is the information gain, whereas CART uses a measure of diversity, for example, the sum of squared errors for normally distributed data. This approach is motivated by the idea of likelihood-maximization. Survival trees extend this approach to exponentially distributed data with censored observations. We used the publicly available R implementation of *rpart* for the analysis.

6. RESULTS

The *k*-NN and the PNN implemented various distance metrics, including the fractal distance [Aggarwal et al., 2001], which is particularly suitable for high-dimensional data. The distance metric was considered an optimization parameter, i.e. we chose that distance metric that provided for the lowest training error rate. The support vector machines implemented three different kernels: linear, radial, and polynomial. The topology of the MLP consisted of one hidden layer with five neurons; the training algorithm was backpropagation (with momentum and adaptive learning rate).

The SVM, PNN, and *k*-NN models were trained in leave-one-out cross-validation (LOOCV) on the learning set. Those model parameters that resulted in the smallest LOOCV error were used for the final model to predict the test cases. The decision tree C5.0 was trained in a 10-fold cross-validation procedure: the learning set was 10 times randomly split into a training set (~70%) and a validation set (~30%). The decision tree used the training set to generate the classification rules and applied them to the validation cases. The tree was pruned in such a way that the classification error on the validation set was minimal. The MLP was trained on the learning set until the accuracy reached 90% or the maximum number of epochs (10,000) was reached. The resulting model was then applied to the test set. Table 1 summarizes the classification results.

Table 1. Correct classification rates in % of the models for the learning (*L*) and test sets (*T*).

Model	*Patient*		*Expression*		*Patient+Expression*		*Tumor Stage+Top Genes*	
	L	*T*	*L*	*T*	*L*	*T*	*L*	*T*
C5.0	70.8	77.5	57.2	67.5	80.2	67.5	80.2	67.5
SVM	74.0	65.0	65.6	65.0	65.6	65.0	74.0	75.0
MLP	89.6	65.0	65.6	72.5	65.6	65.0	88.0	65.0
PNN	73.0	72.5	64.6	62.5	64.6	62.5	71.9	60.0
k-NN	74.0	65.0	65.6	72.5	65.6	72.5	79.2	72.5

The decision tree C5.0 achieved the overall best test set accuracy of 77.5% using the *Patient Data* only. The average accuracy in the learning phase in 10-fold cross-validation was 70.8% (with a standard error of 3.1). Using the expression data only, the average accuracy in the learning phase was 57.2% (with a standard error of 5.6). Using both *Patient+Expression Data,* the tree was able to achieve a higher accuracy in the learning phase (80.2%, standard error of 4.0), but the test accuracy decreased to 67.5%, which might be explained by an overfitting effect.

The PNN achieved the second best performance on the *Patient Data* with a test accuracy of 72.5%. Using the *Expression Data* or *Patient+Expression Data,* the model's performance decreased to 62.5%. The PNN in this study is not able to ignore irrelevant or redundant features, which might explain this degradation. To assess the significance of the result on the *Patient Data,* we performed a random permutation test. In this test, we randomly permuted the class label (i.e., the risk group) of each patient in the learning set, and trained the classifier again. This procedure was repeated 10,000 times to obtain the distribution of the correct classifications under the null hypothesis of random gene expression profiles. The *p*-value ($p = 0.0009$) for the result of the PNN is statistically significant.

The decision tree identified the tumor stage and a set of nine genes *(Top Genes)* as the most relevant variables (data not shown). Based on these variables, the SVM with radial kernel achieved a test set accuracy of 75.0%. The number of correct classifications in the learning set for the unpermuted class labels is 74.0%; this result is statistically significant ($p = 0.0062$, based on a random permutation test involving 10,000 permutations).

With respect to output interpretability and general usability, we believe that the decision tree is the most suitable model in the present study. Decision trees generate classification rules that are easy to understand for humans. The generated classification rules can provide new insights into the structure of microarray data by describing the interrelation between gene expression with respect to the class variable.

We built a survival tree on the set of 211 patients using the variable tumor stage and the quantile-normalized expression values. For each terminal node of the tree, we performed a Kaplan-Meier analysis and used the median of the survival curve to predict the survival time of the patients in this node. For example, let the median survival time in a terminal node be 17.5 months. Then for each patient who falls into this node, we predicted a death event and assumed a survival time of 17.5 months. If the median survival time did not exist in a terminal node (for example, if the node did not contain any death events), then we assumed that the patients in this node were alive. Pruning involves the removal of terminal nodes in a decision tree and is a method for improving the generalization ability of the model. We pruned the survival trees as follows: If two neighboring terminal nodes would both result in the prediction of death events, then these two nodes were merged. If two neighboring nodes both resulted in the prediction of alive, then these two nodes are merged as well. We did not merge two neighboring nodes if they led to different predictions (i.e., one node results in dead, the other one results in alive). This pruning procedure was repeated until no nodes could be merged anymore. Figure 1 shows the resulting survival tree. Node 21 contains 12 patients of early tumor stages (seven patients of tumor stage IA and five patients of tumor stage IB). The mean survival time in this group is 42.4 months; the median survival time is 34.6 months. Node 17 contains 11 patients of early tumor stages as well (two patients of stage IA and nine patients of stage IB). The mean survival time in this group is 15.8 months with a median of 14.2 months. Both node 21 and node 12 contain only death events. Node 6 contains 15 patients of advanced tumor stage (one patient of stage IIA, six patients of IIB, two patients of IIIA, three patients of IIIB, and three patients of IV). The mean survival time is 68.1 months, and the median does not exist. In this group, we observe only four death events. Node 12 contains seven patients of advanced tumor stage (one patients of stage IIA, two patients of IIB, three patients of IIIA, and one patient of IV). We observe only one death event in this group. The mean survival time is 50.2 months; the median does not exist. Figure 2 depicts the Kaplan-Meier survival curves in the four groups.

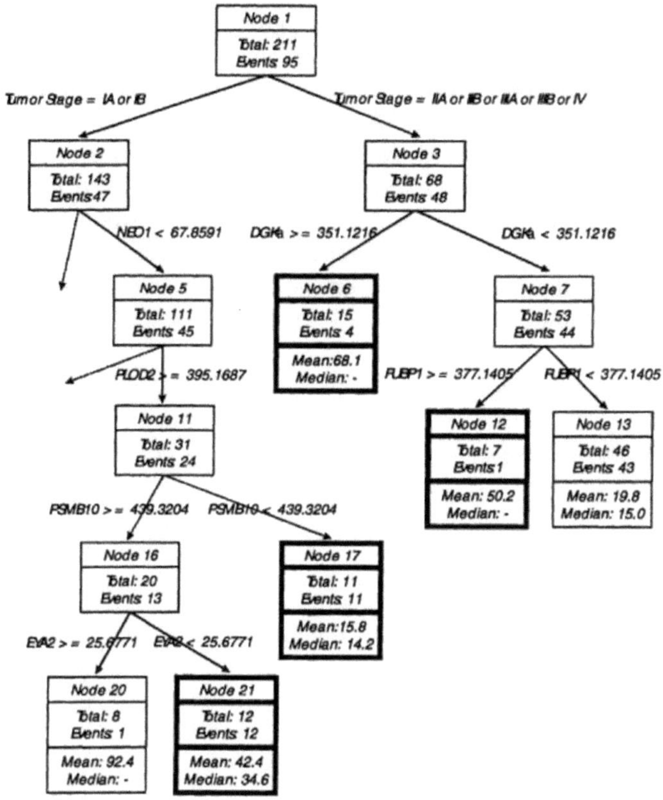

Figure 1. Survival tree (only a small part of the full tree is depicted in the diagram).

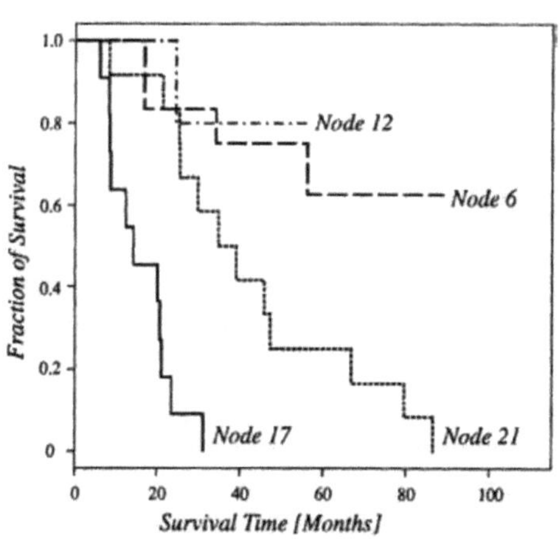

Figure 2. Kaplan-Meier curves of the patients in the nodes 6, 12, 17, and 21.

The survival outcomes in these four groups are surprising, because we would expect that the rather poor outcomes would be associated with advanced tumor stages, while the relatively long survival times would be associated with the early stages. A different distribution of age or sex could possibly explain the observed paradox phenomenon. However, the analysis of the distribution of age and sex in the four groups revealed that neither of these variables is a confounder. Therefore, it can be hypothesized that the surprising outcomes are due to differences in the transcriptional profiles. Table 2 shows the genes that are involved in the groups with unexpected survival outcomes.

Table 2. Genes involved in the groups with unexpected survival outcomes.

Name	*Synonyms, protein names*	*Location*	*GenBank / OMIM no.*	*Key words*
NEO1	Neogenin, NGN	15q22.3-q23	U61262; 601907	Tissue growth regulation, cell-cell recognition, and cell migration
PLOD2	Procollagen-lysine 2-oxoglutarat 5-dioxygenase 2, lysine hydroxylase 2, lysyl hydroxylase 2	3q23-q24	AY026757; 601865	Collagen maturation
PSMB10	Proteasome subunit beta-type 10, MECL1, LMP10, PSBA	16q22.1	AH011134; 176847	Proteolysis and peptidolysis, humoral defense mechanism
EYA2	Homologue of Eyes Absent Drosophila 2, DRES12, EAB1	20q13.1	AF387364; 601654	Development of eye; triggers rapid apoptosis in interleukin-3-dependent 32D.3 murine myeloid cells
DGKα	Diacylglycerol kinase alpha, DAGK, 80-KD	12q13.3	AY335740; 125855	Intracellular signaling pathway
FUBP1	Far upstream element-binding protein 1, FUSE-binding protein, KHSRP, FBP2, KSRP	19p13.3	AH007695; 603444	Neuron-specific splicing of the N1 exon of SRC

7. DISCUSSION

In the following, we briefly discuss the biology of the six genes whose expression profiles might be associated with the outcome of the "surprising" survival groups.

The most important gene is NEO1. The protein encoded by this gene, neogenin, shares 53% amino acid identity with DCC (Deleted in Colorectal Cancer), a candidate tumor suppressor gene [Vielmetter et al., 1997]. DCC is important for the maintenance of normal tissue differentiation in multiple tissues, and its deletion has been shown to result in proliferation of various cancers [Vielmetter et al., 1997]. Based on their sequence conservation and similar expression during development, Meyerhardt et al. [1997] hypothesized that DCC and neogenin have related functions. However, while the loss of DCC expression is frequent in various cancers, the researchers found out that the expression of neogenin is not altered in more than 50 tumor types. Due to its chromosomal location (15q22.3-q23) and ubiquitous expression, neogenin seems to be infrequently altered in cancer. Neogenin is also highly expressed in various embryonic tissues, suggesting a more general role in developmental processes such as tissue growth regulation, cell-cell recognition, and cell migration [Vielmetter et al., 1997]. Although no direct link between neogenin and tumor suppression has been found yet, Vielmetter el al. [1997] hypothesized that NEO1 encodes a regulatory protein in the transition of undifferentiated proliferating cells to their differentiated state. Other recent studies have revealed that the expression of neogenin is associated with esophageal squamous cell carcinoma [Hu et al., 2001] and breast cancer [Srinivasan et al., 2003]. According to the survival tree, underexpression of neogenin is associated with the low survival groups of the early lung cancer patients in node 17 and node 21. If neogenin acts as a tumor suppressor, then the underexpression of neogenin might have an impact on the survival outcome of the patients. It might be possible that some rare aggressive forms of early lung adenocarcinomas are associated with the loss of expression of neogenin, while other, less aggressive subtypes, are not. The question whether the loss of expression of neogenin is causal for the poor survival outcome of the patients in nodes 17 and 21 remains open. However, the results of the survival tree corroborate the hypotheses of Vielmetter et al. [1997] and Meyerhardt et al. [1997]

PLOD2 encodes for Procollagen-lysine 2-oxoglutarat 5-dioxygenase 2 and is located on 3q23-q24; this protein catalyzes the hydroxylation of lysyl residues in collagens [OMIM no. 601865]. Denko et al. [2003] observed an overexpression of PLOD2 in hypoxic epithelial cells. The researchers hypothesized that the differential expression of PLOD2 could contribute to hypoxia-induced metastasis. Hypoxia is known to be a potent factor for tumor angiogenesis in lung adenocarcinomas [Sato et al., 2002]. The survival tree suggests that overexpression of PLOD2 correlates with a poor clinical outcome for patients of early tumor stages.

PSMB10 (aka LMP10) is located on 16q22.1 and is involved in proteolysis and peptidolysis, the humoral defense mechanism, and in an ATP/ubiquitin-dependent non-lysosomal proteolytic pathway [GenAtlas, 2003]. Some cancerous cells are able to alter the expression of proteins that are involved in antigen processing, so that cytoxic T-cells do not recognize the tumor. Using human cancer cell lines, Johnsen et al. [1998] investigated multiple genes that are differentially expressed in the class I MHC antigen-processing pathway, including several proteasome subunits that have been implicated in antigen processing. They observed a complete loss of expression of TAP1, TAP2, LMP2, and LMP7, as well as PSMB10, which is encoded outside the MHC pathway. Johnsen et al. [1998] hypothesized that some tumors may alter the immune surveillance by simultaneously down-regulating multiple components of the MHC-I antigen-processing pathway, which results in an alteration of the processing and presentation of tumor antigens. According to the results of the survival tree, underexpression of PSMB10 is associated with a very poor survival outcome (cf. node 17). Some aggressive types of lung adenocarcinomas might suppress the expression of PSMB10 and thereby altering the MHC antigen-processing pathway, so that cytoxic T-cells are no longer able to recognize the cancer cells. This "camouflage effect" is not observed in the case that PSMB10 is not underexpressed, leaving the possibility of a better survival outcome (cf. node 20). Recently, Huang et al. [2003] have identified PSMB10 as a gene associated with metagene predictors of breast cancer recurrence.

EYA2 is located on 20q13.1 and is the human homologue of the Eyes Absent gene in *Drosophila.* This gene plays a pivotal role in the development of the *Drosophila* eye; without this gene, progenitor cells in the eye imaginal disc undergo programmed cell death [Clark et al., 2002]. If EYA2 has a functional homology to *Drosophila* EYA, then it may be involved in apoptosis as well. Clark et al. [2002] recently reported that a misexpression of members of the Eyes Absent family triggers apoptosis. According to the results of the survival tree in our analysis, the expression of EYA2 is crucial for dividing the patients into a group of good clinical outcome (node 20), and the group of poor clinical outcome (node 21).

Some patients suffering from an advanced type of lung adenocarcinoma have a surprisingly good clinical outcome, e.g. the patients in node 6. An overexpression of DGKα is associated with this group. The average survival time of these patients is over 5 years. DGKα is Diacylglycerol kinase α, and is located on 12q13.3. Diacylglycerol (DAG) functions in intracellular signaling pathways as an allosteric activator of protein kinase C [OMIM no. 125855]. Furthermore, DAG appears to be involved in regulating RAS family proteins [OMIM no. 125855]. Topham et al. [2001] reported that the regulation of DAG is crucial to maintain cellular homeostasis. DAG kinases phosphorylate DAG to phosphatidic acid (PA), thereby suppressing the function of DAG. In cancer cells, DAG is often overexpressed, and also PA can lead to abnormal cell division and cancer. Apparently, the tight regulation of these kinases is crucial for the normal cell development. In their investigation, Topham et al. [2001] focused on the impact of DGKζ on the regulation of the gene Ras. Guanine nucleotide exchange factors (GEFs) activate Ras by facilitating GTP binding. RasGRP, an exchange factor, was recently identified as a potential leukemia disease gene. An overexpression of this protein in cultured cells leads to a transformed phenotype. According to Topham et al. [2001], these observations indicate that abnormally high RasGRP activity can lead to malignant transformation. RasGRP has a diacylglycerol (DAG)-binding domain, and its activity as exchange factor depends on local concentration of DAG. Since DAG kinases remove DAG from the cell by converting it to PA, they are able to reduce the concentration of DAG. Consequently, these kinases might serve as an "off-mechanism" for RasGRP, and thereby reduce the activation of Ras. DGK kinases may play a pivotal role in the Ras signaling pathway. However, Topham et al. found that only one DGK isoform, namely DGKζ, is able to affect RasGRP activity significantly. The survival tree in our analysis identified the expression level of DGKα as an important discriminator for the patients with advanced tumor stages: whereas an overexpression of DGKα is associated with a rather good clinical outcome, an underexpression is associated with a rather poor outcome.

Most patients suffering from an advanced tumor stage and showing an underexpression of DGKα have a rather poor clinical outcome (cf. node 7, containing 53 patients and 44 death events). However, for some of these patients, we observe an overexpression of FUBP1. These patients have a remarkably better survival outcome (cf. node 12). FUBP1 is a fuse-binding protein, and the encoding gene is located on 1p31.1 [OMIM no. 603444]. FUBP1 is a transcriptional activator of c-myc [Kim et al., 2003]. Kim et al. [2003] have recently shown that overexpression of c-myc is frequently associated with cancers in various tissues and organs, including lung, and its expression is suppressed during lung differentiation. Binding of the tRNA synthetase cofactor p38 stimulates ubiquitination and degradation of FUBP1, leading to downregulation of myc, which is required for differentiation of functional alveolar type II cells. Adenocarcinomas are known to arise in distal portions of the airway and alveolus. [Borczuk et al., 2003]. c-myc is known to play a pivotal role in cell growth, and an overexpression of c-myc is oncogenetic [Takahashi et al., 1998; He et al., 2000]. The tight regulation of c-myc is crucial for normal cell growth, and the expression of FBP1 at a proper level is therefore required. Liu et al. [2001] have shown that mutations of the TFIIH helicase that impair regulation by FBP1 affect proper regulation of c-myc expression and have implications in the development of malignancy. Interestingly, Borczuk et al. [2003] have recently identified FUBP1 as one of the top-100 marker genes in large cell lung carcinoma.

The survival outcome of lung cancer patients depends from various different factors and is certainly very difficult to predict. In the present study, integrating clinical and transcriptional data did not result in an improved prediction of the 5-year survival outcome. Here, the tumor stage is the most important predictor.

However, the regression study revealed that gene expression profiling of cancer specimens might contain some information about the clinical course. Using survival trees as exploratory tools rather than prediction models, it was possible to gain new insights in the structure of the data set. Whether the identified genes play a key role in the clinical course of lung adenocarcinoma patients remains an open question and requires adequate validation by molecular experiments. But the present study illustrated how survival trees could be used as exploratory tools, similarly to hierarchical clustering approaches that are already widely used to structure and visualize microarray data.

REFERENCES

Aggarwal CC, Hinneburg A, Keim DA (2001) On the surprising behavior of distance metrics in high dimensional space. Proc 8th Inter Conf Database Theory (ICDT), 420-434.

Beer DG, Kardia SL, Huang CC, Giordano TJ, Levin AM, Misek DE, Lin L, Chen G, Gharib TG, Thomas DG, Lizyness ML, Kuick R, Hayasaka S, Taylor JM, Iannettoni MD, Orringer MB, Hanash S (2002) Gene-expression profiles predict survival of patients with lung adenocarcinoma. Nat Med 8(8):816-24.

Berrar D, Downes CS, Dubitzky W (2003), Multiclass cancer classification using gene expression profiling and probabilistic neural networks. Proc Pac Symp Biocomp 8:5-16.

Bhattacharjee A, Richards WG, Staunton J, Li C, Monti S, Vasa P, Ladd C, Beheshti J, Bueno R, Gillette M, Loda M, Weber G, Mark EJ, Lander ES, Wong W, Johnson BE, Golub TR, Sugarbaker DJ, Meyerson M (2001) Classification of human lung carcinomas by mRNA expression profiling reveals distinct adenocarcinoma subclasses. Proc Natl Acad Sci USA 98(24):13790-13795.

Bolstad B.M., Irizarry R, Astrand M, Speed TP (2002) A comparison of normalization methods for high density oligonucleotide array data based on variance and bias. Bioinformatics 19(2):185-93.

Borczuk AC, Gorenstein L, Walter KL, Assaad AA, Wang L, Powell CA (2003) Non-small-cell lung cancer molecular signatures recapitulate lung developmental pathways. Am J Path 163(5): 1949-1960.

Breiman L, Friedman J, Olshen RA, Stone CJ (1984) Classification and Regression Trees. Chapman & Hall, New York.

Brown MPS. Grundy W, Lin D, Cristianini N, Sugnet C, Furey T, Ares M, Jr., Haussler D (2000) Knowledge-based analysis of microarray gene expression data by using support vector machines. Proc Natl Acad Sci USA. 97(1):263-267.

Burges C (1998) A tutorial on support vector machines for pattern recognition. Data Mining and Knowledge Discovery 2(2).

Clark SW, Fee BE, Cleveland JL (2002) Misexpression of the eyes absent family triggers the apoptotic program. J Biol Chem 277(5):3560-3567.

Denko NC, Fontana LA, Hudson KM, Sutphin PD, Raychaudhuri S, Altman R, Giaccia AJ (2003) Investigating hypoxic tumor physiology through gene expression patterns. Oncogene 22:5907-5914.

GenAtlas(May 1, 2003), http://www.dsi.univ-paris5.fr/genatlas/fiche.php?symbol=PSMB10.

He L, Liu J, Collins I, Sanford S, O'Connell B, Benham CJ, Levens D (2000) Loss of FBP function arrests cellular proliferation and extinguishes c-myc expression. EMBO J 19(5): 1034-1044.

Hu YC, Lam KY, Law S, Wong J, Srivastava G (2001) Identification of differentially expressed genes in esophageal squamous cell carcinoma (ESCC) by cDNA expression array: overexpression of Fra-l, neogenin, Id-l, and CDC25B genes in ESCC. Clin Cancer Res 2213(7):2213-2221.

Huang E, Cheng SH, Dressman H, Pittman J, Tsou MH, Horng CF, Bild A, Iversen ES, Liao M, Chen CM, West M, Nevins JR, Huang AT (2003) Gene expression predictors of breast cancer outcomes. Lancet 361(9369):1590-1596.

Johnsen A, France J, Sy MS, Harding CV (1998) Down-regulation of the transporter for antigen presentation, proteasome subunits, and class I major histocompatibility complex in tumor cell lines. Cancer Res 58(16):3660-3667.

Khan J, Wei JS, Ringnér M, Saal LH, Ladanyi M, Westermann F, Berthold F, Schwab M, Antonescu CR, Peterson C, Meltzer PS (2001) Classification and diagnostic prediction of cancers using gene expression profiling and artificial neural networks. Nat Med 7(6):673-679.

Kim MJ, Park BJ, Kang YS., Kim HJ, Park JH, Kang JW, Lee SW, Han JM, Lee HW, Kim S (2003) Downregulation of FUSE-binding protein and c-myc by tRNA synthetase cofactor p38 is required for lung cell differentiation. Nat Gen 34:330-336.

Liu J, Akoulitchev S, Weber A, Ge H, Chuikov S, Libutti D, Wang XW, Conaway JW, Harris CC, Conaway RC, Reinberg D, Levens D (2001) Defective interplay of activators and repressors with TFIH in xeroderma pigmentosum. Cell 104(3):353-63.

Meyerhardt JA, Look AT, Bigner SH, Fearon ER (1997) Identification and characterization of neogenin, a DCC-related gene. Oncogene 14(10): 1129-1136.

Ochs MF, Godwin AK (2003) Microarrays in cancer: research and applications. BioTechniques 34: pp. S4-S15.

OMIM, (May 1, 2004), http://www.ncbi.nlm.nih.gov/Omim/. *The # refers to the database entry.*

Sato M, Tanaka T, Maeno T, Sando Y, Suga T, Maeno Y, Sato H, Nagai R, Kurabayashi M (2002) Inducible Expression of Endothelial PAS Domain Protein-1 by Hypoxia in Human Lung Adenocarcinoma A549 Cells. Am J Resp Cell Mol Biol 26(1): 127-134.

Shipp MA Ross KN, Tamayo P, Weng AP, Kutok JL, Aguiar RCT, Gaasenbeek M, Angelo M, Reich M, Pinkus GS, Ray TS, Koval MA, Last KW, Norton A, Lister TA, Mesirov J, Neuberg DS, Lander ES, Aster JC, Golub TR (2002) Diffuse large B-cell lymphoma outcome prediction by gene-expression profiling and supervised machine learning. Nat Med 8:68-74.

Specht DF(1990) Probabilistic neural networks. Neural Networks 3:109-118.

Srinivasan K, Strickland P, Valdes A, Shin GC, Hinck L (2003) Netrin-1/neogenin interaction stabilizes multipotent progenitor cap cells during mammary gland morphogenesis. Devel. Cell 4(3):371-82.

Takahashi T, Konishi H, Kozaki K, Osada H, Saji S, Takahashi T, Takahashi T (1998) Molecular analysis of a myc antagonist, ROX/Mnt, at 17pl3.3 in human lung cancers. Jap J Cancer Res 89:347-351.

Topham MK, Prescot SM (2001) Diacylglycerol kinase zeta regulates Ras activation by a novel mechanism. JCellBiol 152:1135-1143.

van't Veer LJ, Dai HY, van de Vijver MJ, He YDD, Hart AAM, Mao M, Peterse HL, van der Kooy K, Marton MJ, Witteveen AT, Schreiber GJ, Kerkhoven RM, Roberts C, Linsley PS, Bernards R, Friend SH (2002) Gene expression profiling predicts clinical outcome of breast cancer. Nature 415:530-536.

Vielmetter J, Chen XN, Miskevich F, Lane RP, Yamakawa K, Korenberg JR, Dreyer WJ (1997) Molecular characterization of human neogenin, a DCC-related protein, and the mapping of its gene(NEO1) to chromosomal position 15q22.3-q23. Genomics 41(3):414-421.

Wang M, Lemon WJ, Liu G, Wang Y, Iraqi FA, Malkinson AM, You M (2003) Fine mapping and identification of candidate pulmonary adenoma susceptibility genes using advanced intercross lines. Cancer Res 63:3317-3324.

Yeoh EJ, Ross ME, Shurtleff SA, Williams WK, Patel D, Mahfouz R, Behm FG, Raimondi SC, Relling MV, Patel A, Cheng C, Campana D, Wilkins D, Zhou X, Li J, Liu H, Pui CH, Evans WE, Naeve C, Wong L, Downing JR (2002) Classification, subtype discovery, and prediction of outcome in pediatric acute lymphoblastic leukemia by gene expression profiling. Cancer Cell 1:133-143.

Zhang H, Yu CH, Singer B, Xiong M (2001) Recursive partitioning for tumor classification with gene expression microarray data. Proc Natl Acad Sci USA 98(12):6730-6735.

Chapter 12

USE OF MICRO ARRAY DATA VIA MODEL-BASED CLASSIFICATION IN THE STUDY AND PREDICTION OF SURVIVAL FROM LUNG CANCER

Liat Ben-Tovim Jones[1,2], Shu-Kay Ng[1], Christophe Ambroise[3], Katrina Monico[1], Nazim Khan[1] and Geoff McLachlan[1,2]
[1]Department of Mathematics and [2]Institute for Molecular Bioscience, University of Queensland, Brisbane, Australia; [3]Laboratoire Heudiasyc, Centre National de la Recherche Scientifique, Compiegne, France

Abstract: We applied a model-based clustering approach to classify tumor tissues on the basis of microarray gene expression. The impact of this classification on cancer biology and clinical outcome was studied. In particular, the association between the clusters so formed and patient survival (recurrence) times was examined. The approach was illustrated using the four CAMDA'03 lung cancer datasets. We showed that the gene expression-based clustering is a powerful predictor of the outcome of disease, in addition to current systems based on histopathology criteria and extent of disease at presentation.

Key words: Mixture models, EMMIX-GENE algorithm, selection bias, microarrays, survival analysis, Cox proportional hazards, Kaplan-Meier survival curve

1. INTRODUCTION

Lung cancer patients with the same stage of disease can have markedly different treatment responses and clinical outcome. Recent studies have suggested that information from gene expression profiles can be used to classify cancer tissues by type and subtype; see, for example, Mateos et al. [2001] and Wigle et al. [2002]. The aim of our analysis was to demonstrate that the gene expression data provided additional information on survival beyond that provided by the histopathology of the tumors. We applied a model-based approach to cluster tumor tissues on the basis of gene expression, implemented in the EMMIX-GENE algorithm [McLachlan et al., 2002]. The impact of this clustering of tissues on cancer biology and clinical outcome was investigated. The association between clusters formed and patient survival (or recurrence) times was examined. We analyzed each of the four CAMDA'03 datasets, referred to here as Ontario [Wigle et al., 2002], Stanford [Garber et al., 2001], Harvard [Bhattacharjee et al., 2001] and Michigan [Beer et al., 2002] datasets. The first two used cDNA arrays, while the latter two used different versions of Affymetrix oligonucleotide arrays. We analyzed each dataset individually in order to determine whether we could make conclusions from a particular dataset in its own right.

2. ANALYTICAL METHODS

2.1 Data Selection

We downloaded the processed data from the CAMDA'03 contest web site (http://www.camda.duke.edu/camda03). For the cDNA arrays, we took the 2880 (Ontario dataset) and 918 (Stanford dataset) genes. For the Affymetrix arrays, we started with the 3312 (Harvard dataset) and 4965 (Michigan dataset) genes, but these had outlier values. Thus, for the Harvard dataset, we imposed a floor (lower bound) of 1 and a ceiling (upper bound) of 3000 (leaving 3190 genes), and the data were then log transformed. For the Michigan dataset, we imposed a floor of –1 and a ceiling of 26,000 (reducing to 4728 genes) and then applied the generalized log transformation, $\log(x+\sqrt{c^2+x^2})$. The microarray data matrix for each dataset was formed where each row of the data matrix represents a single gene and each column a tumor tissue. Each column of the data matrix was standardized to have mean zero and unit standard deviation. Finally, each row of the consequent matrix was standardized to have mean zero and unit standard deviation. In all the datasets, we imputed missing values using the method in Dudoit et al. [2002]. By comparing Unigene identifiers, we found that the cDNA arrays had at least 105 genes in common, while the Affymetrix arrays had at least 1257 genes in common in the input datasets.

2.2 Model-Based Clustering Approach

The EMMIX-GENE algorithm has been developed for the specific purpose of the clustering of tissue samples on a very large number of genes [McLachlan et al., 2002]. The first step of the algorithm considered the selection of a subset of relevant genes from the available set of genes. This selection process was undertaken in the absence of tissue samples that were of known classification with respect to the clinical outcome. Briefly, we tested for each gene separately whether there was genuine grouping in the tissues. Based on the likelihood ratio test statistic, relevant genes that revealed the group structure for clustering tissue samples were retained. In the second step, we clustered the retained genes into a user-specified number of groups so that highly correlated genes were placed in the same cluster. Each gene-cluster (metagene) was represented by the sample mean of the cluster. The final step concerned the clustering of the tissues by fitting mixtures of factor analyzers [McLachlan et al., 2000]. It was undertaken on the basis of the metagenes [McLachlan et al., 2002].

2.3 Survival Analysis

In the survival analysis, only tissues from patients with known clinical characteristics and survival times were considered. With the Ontario dataset, we defined the outcome as the time between surgery and the recurrence. This is equivalent to the definition in Wigle et al. [2002] because patients free from recurrence are all still alive at the end of follow-up period. There were 37 patients with 15 censored. For the Stanford dataset, there were 26 Adenocarcinoma (AC) tissues with four tumor pairs derived from the same patients. In the analysis, each tumor pair was treated as one observation. This gave 22 observations in total, with 10 censored. With the Harvard dataset, there were 115 patients and 64 died before the end of the follow-up. For the Michigan dataset, there were 86 patients and 24 died before the end of the follow-up.

The Kaplan-Meier method was used to estimate the overall survival (or being recurrence-free) of patients for each cluster formed by the gene expression-based clustering. Kaplan-Meier survival curves of the clusters were compared using the log-rank test. The impact of the gene expression-based clustering of tissues on patient survival was studied using the Cox proportional hazards model [Cox, 1972]. It was determined by examining the relative hazard ratios with respect to the clusters of the tissues. The significance of estimated hazard ratios were tested using the Wald test. A significant result implies that clinical outcomes on the basis of patient survival are different between clusters. All calculations were performed with the S Plus statistical package.

3. RESULTS

3.1 Clustering of Tumor Tissues

We ran EMMIX-GENE for all the tumor types within each dataset. We retrieved the histological classification for at least the non-AC tumors in all datasets, except for the Ontario dataset. For the other three datasets we focused on the AC tumors, since the clinical data were available only for this type of tumor. Each row of the reduced data matrix was then standardized to have mean zero and unit standard deviation. We re-ran EMMIX-GENE on the reduced set of AC tumors only.

For the Ontario dataset, we retained 766 genes. The top genes (as indicated by the likelihood ratio statistic) included immunoglobulin lambda light chain IGL, hypothetical protein FLJ10404, HLA-B associated transcript 2 D6S51E, Friend leukemia virus integration 1 FLI1 and ATP-binding cassette ABCD3. The heat maps (not given here) were adopted to exhibit similarities between clusters of the tissue samples. They present a grid of colored points where each color represents a gene-expression value for a gene in the tissue sample. It was found that several metagenes clearly separated tissues into two clusters, which we termed poor- and good-prognosis clusters. The former cluster comprised 23 of the 24 patients with recurrence, plus eight patients with censored survival data, suggesting that these might have poorer outcome. Wigle et al. [2002] also found five of these in their "early recurrence" cluster. The good-prognosis cluster comprised the remaining seven recurrence-free patients, and also a patient known to recur (P171 ADC). Wigle et al. [2002] also found this, and this patient was still alive at the end of the follow-up.

For the Stanford dataset, we clustered a subset of 35 AC tumors (we removed the tumors which classified into non-AC clusters when clustering the full dataset). We retained 219 genes, of which the top genes included CD36 antigen, signal transducer and activator of transcription 4, aldo-keto reductase family 1 member C1 and kynureninase. We clustered the tissues into two clusters, corresponding to the poor-prognosis and good-prognosis clusters. Our poor-prognosis cluster corresponded to the Garber et al. [2001] AC group 3 (worst clinical outcome), while the good-prognosis cluster corresponded to AC groups 1 and 2. The only exception was tissue from patient 218 (AC group 3), which appeared in our good-prognosis cluster.

For the Harvard dataset, we clustered 127 AC tumors and retained 858 genes. The top five genes were: tubulin-specific chaperone e, thioredoxin reductase 1, UDP-glucose dehydrogenase, Cluster Incl AL096723 and the gene protein kinase, interferon-inducible double stranded RNA dependent. We obtained three AC clusters, which did not appear to correspond to those of Bhattacharjee et al. [2001] (AC subtypes C1-C4), but rather were spread throughout their clusters.

For the Michigan dataset, we clustered 86 AC tumors and retained 1394 genes. The top genes were: MUC3A (mucin 3), TRAP1 (heat shock protein 75), HLA-DQA1 (major histocompatibility complex class II), RPS4Y (human ribosomal protein) and POU2AF1 (POU domain class 2 associating factor). We obtained three tissue clusters, and these overlapped with those of Beer et al. [2002], with 25 tissues differing between our results and theirs.

3.2 Identification of Prognostic Genes for AC Tumors

We wanted to see if we could unify our AC tumor clusters, by finding common genes important in identifying clusters between the datasets. For the Affymetrix arrays (Harvard and Michigan datasets), the EMMIX-GENE procedure retained at least 108 genes common to both. (These included interesting genes involved in metabolism, transcription and translation, cell signaling and cell cycle control.) We matched these by gene name to the retained genes (219) with the Stanford dataset. We found at least six common genes, four of which are involved in cellular metabolism (thioredoxin reductase 1, ornithine decarboxylase 1, S100 calcium-binding protein A10 and kynureninase). We also identified full-length proliferating cell nuclear antigen (PCNA), a DNA binding protein involved in control of replication and epithelial membrane protein 2 (EMP2), a reported tumor-associated gene. Several of the metabolic enzymes were also found to be important in the original papers on these datasets, for example kynureninase appeared as a top gene in our retained genes (Stanford dataset) and was found by Beer et al. [2002] in the top 100 marker genes. Also, Garber et al. [2001] found ornithine decarboxylase and thioredoxin reductase as markers to differentiate their long-term survivor and poor-prognosis groups. PCNA was mentioned in Bhattacharjee et al. [2001] as a marker gene for cluster C1, though this cluster was not associated in their study with clinical outcome.

3.3 Impact of Classification on Outcomes

For the Ontario dataset, the Kaplan-Meier curves (Figure 1) showed a significant difference in the probability of recurrence-free survival between the good-prognosis and poor-prognosis clusters (p-value=0.027). The mean (±SE) times between surgery and recurrence were 1388±156 and 665±86 days, respectively. The results of the multivariate Cox regression analysis are given in Table 1. The gene expression-based cluster indicator variable was the only factor near to significance at the conventional 5% level (p-value=0.06).

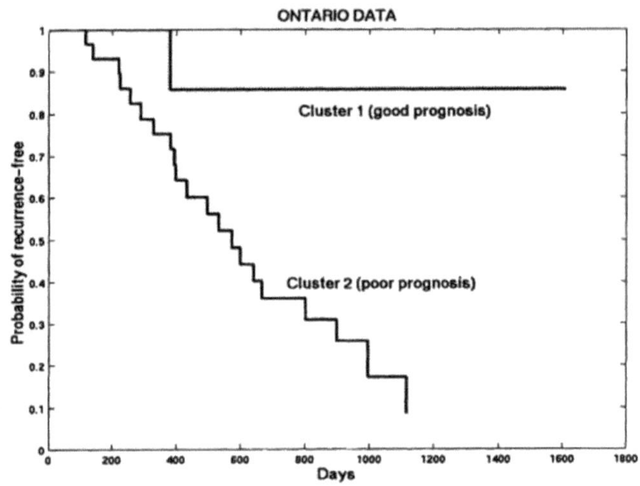

Figure 1. Kaplan-Meier curves of recurrence-free survival for the two clusters (Ontario data).

Table 1. Multivariate Cox hazards analysis of the risk of recurrence (Ontario data).

Variable	Hazard ratio (95%CI)	p-value
Poor-prognosis cluster (vs. good prognosis)	6.8 (0.9-51.8)	0.06
Stages 2 or 3 (vs. Stage 1)	1.1 (0.4-2.7)	0.88

For the Stanford dataset, the Kaplan-Meier survival curves (Figure 2) showed a significant difference in the probability of overall survival between the good-prognosis and poor-prognosis clusters (p-value<0.001). The mean (±SE) survival times were 37.5±5.0 and 5.2±2.3 months, respectively. The results of the multivariate Cox regression analysis are given in Table 2. It was evident that the two prognosis clusters were different after the adjustment for the clinical factors (p-value=0.002). The estimated hazard ratio for overall survival in the poor-prognosis cluster as compared with the good-prognosis cluster was 15.5 (95% CI: 2.7 to 90.2).

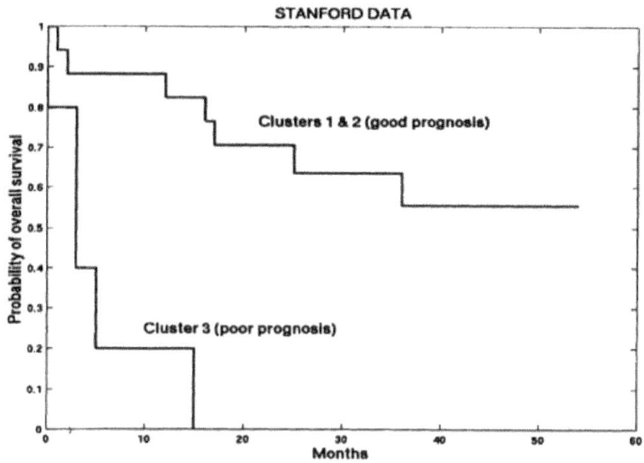

Figure 2. Kaplan-Meier survival curves for the two prognosis clusters (Stanford data).

Table 2. Multivariate Cox proportional hazards analysis of the risk of death (Stanford data).

Variable	Hazard ratio (95%CI)	*p*-value
Poor-prognosis cluster (vs. good-prognosis)	15.5 (2.7–90.2)	0.002
Tumor grade 3 (vs. grades 1 or 2)	1.8 (0.4 – 9.2)	0.47
Tumor size 1 (vs. sizes 2 to 4)	0.5 (0.03–7.4)	0.59
Presence of tumor in lymph nodes	4.4 (0.4–48.6)	0.23
Presence of metastases	4.3 (0.8–24.6)	0.10

With the Harvard dataset, the mean (±SE) survival times were 62.2±5.7, 50.9±5.8, and 26.5±4.7 months for Clusters 1 to 3, respectively. The Kaplan-Meier survival curves for the three clusters are displayed in Figure 3. They indicated that survival in Cluster 3 differed significantly relative to that in Cluster 1 (*p*-value=0.014) and in Clusters 1 and 2 combined (*p*-value=0.043).

The gene expression-based clustering was significantly associated with the clinical outcome on the basis of survival. The results of the multivariate Cox regression analysis (the variable presence of metastases was not included in the analysis due to too many missing data) are shown in Table 3. It was found that the patient survival for Cluster 3 was different from Clusters 1 and 2 combined, after adjustment for the clinical factors (*p*-value=0.008).

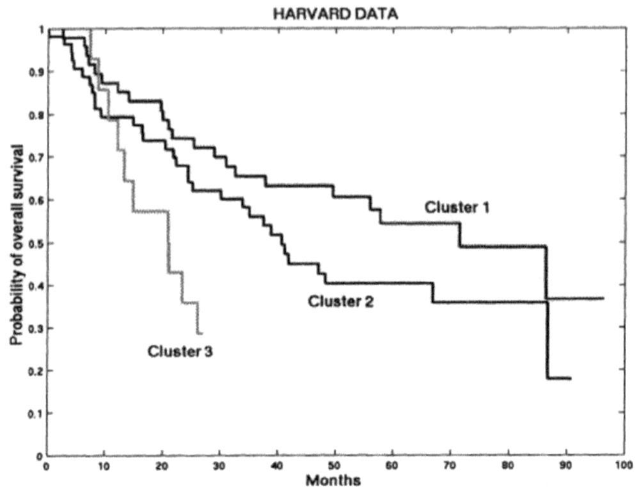

Figure 3. Kaplan-Meier survival curves for the three clusters (Harvard data).

Table 3. Multivariate Cox proportional hazards analysis of the risk of death (Harvard data).

Variable	Hazard ratio (95%CI)	*p*-value
Cluster 3 (vs. clusters 1 or 2)	2.7 (1.3–5.4)	0.008
Age	1.0 (1.0–1.1)	0.086
Female vs. Male	0.6 (0.3–1.1)	0.078
Smoking frequency	1.3 (0.6–2.5)	0.520
Tumor size 1 (vs. sizes 2 to 4)	1.7 (0.9–3.3)	0.110
Presence of tumor in lymph nodes	2.4 (1.3–4.7)	0.009
Grade 1 (vs. grades 2 to 4)	1.5 (0.7–3.1)	0.260

For the Michigan dataset, the mean (±SE) survival times were 86.2±7.1, 60.3±8.0, and 48.0±7.9 months for Clusters 1 to 3, respectively. The Kaplan-Meier survival curves presented in Figure 4 showed that survival in Cluster 1 differed relative to that in Cluster 2 (*p*-value=0.056) and approached significance for Clusters 2 and 3 combined (*p*-value=0.069).

The results of the multivariate Cox regression analysis are given in Table 4. As the tumor stage and the number of tumors in lymph nodes were highly correlated (Spearman's rank correlation: 0.93), only the former was included in the analysis. It can be seen that the tumor stage was the only significant factor affecting survival.

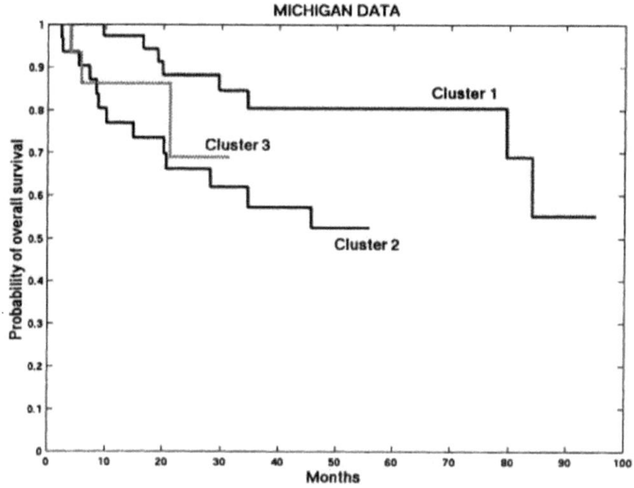

Figure 4. Kaplan-Meier survival curves for the three clusters (Michigan data).

Table 4. Multivariate Cox proportional hazards analysis of the risk of death (Michigan data).

Variable	Hazard ratio (95%CI)	*p*-value
Clusters 2 or 3 (vs. Cluster 1)	1.8 (0.7– 4.5)	0.220
Age	1.1 (1.0– 1.1)	0.064
Female (vs. Male)	0.5 (0.2– 1.3)	0.160
Stage 3 (vs. stage 1)	5.3 (1.8–16.1)	0.003
Tumor sizes 3 or 4 (vs. sizes 1 or 2)	1.6 (0.4– 6.5)	0.510
Moderate or Poor differentiation (vs. well)	1.4 (0.3– 5.5)	0.660

4. DISCUSSION

We used a mixture-model based approach to identify patient clusters. The advantage of this method is that it provided a sound statistical basis for clustering and for assessing what is the right number of clusters. Our analysis was limited by the small numbers of tumors available (especially for the Ontario and Stanford datasets). In addition, clinical data were available for only subsets of the tumors (often only for one tumor type, AC), and the high proportion of censored observations limited the comparison of survival curves among clusters (for example, with the Michigan dataset, the percentage of censored observations was 72%).

We focused here on the use of cluster analysis (unsupervised classification) to link gene-expression data with survival. We also used discriminant analysis (supervised classification) to illustrate the prognostic value of the clusters obtained on the basis of the gene expressions. We proceeded on the basis that the clusters so formed represented training data of known origin from the classes with varying degrees of survival and formed a support vector machine (SVM) for the prediction of the class of origin of a new tumor. This was somewhat self-serving in taking the clusters to be random samples from the underlying prognostic classes, but it at least provided a lower bound on the error rate that could be expected of a prediction rule based on genuine training data. We used the SVM with recursive feature elimination (RFE) of Guyon et al. [2002J to eliminate genes in a backward selection procedure from the SVM, using (10-fold) cross-validation [Ambroise and McLachlan, 2002] with allowance for the selection bias. It was found that the error in predicting the clinical outcome of a new lung cancer tumor was approximately 6% (Ontario dataset), 3% (Stanford dataset), 5% (Harvard dataset), and 26% (Michigan dataset). Also, the prediction rule provided a method for revealing potential marker genes, as it can be noted how many times a gene was selected in the final form of the SVM on each of the ten sub-samples during the 10-fold cross-validation.

5. CONCLUSIONS

We applied a model-based clustering approach to classify tumor tissues using their gene signatures into (a) clusters corresponding to tumor type and (b) clusters corresponding to clinical outcomes for tumors of a given subtype. In (a) we found almost perfect correspondence between cluster and tumor type, at least for non-AC tumors, except in the Ontario dataset. The clusters in (b) were identified with clinical outcomes such as recurrence versus non-recurrence and death versus long-term survival. Except for the Michigan dataset, we were able to show that gene expression data provided prognostic information, beyond clinical indicators such as stage.

6. ACKNOWLEDGEMENTS

We thank Richard Bean, Abdollah Khodkar and Justin Zhu for their assistance with this paper.

REFERENCES

Ambroise, C. and McLachlan, G.J. (2002). Selection bias in gene extraction on basis of microarray gene expression data. *Proc. Natl. Acad. Sci. USA,* 99, 6562-6566.

Beer, DJ, DG, Kardia SL, Huang CC, Giordano TJ, Levin AM, Misek DE, Lin L, Chen G, Gharib TG, Thomas DG, Lizyness ML, Kuick R, Hayasaka S, Taylor JM, lannettoni MD, Orringer MB, Hanash S. (2002). Gene-expression profiles predict survival of patients with lung adenocarcinoma. *Nature Medicine,* 8, 816-824.

Bhattacharjee, A, Richards WG, Staunton J, Li C, Monti S, Vasa P, Ladd C, Beheshti J, Bueno R, Gillette M, Loda M, Weber G, Mark EJ, Lander ES, Wong W, Johnson BE, Golub TR, Sugarbaker DJ, Meyerson M (2001). Classification of human lung carcinomas by mRNA expression profiling reveals distinct adenocarcinoma subclasses. *Proc. Natl. Acad. Sci.* (*USA*, 98, 13790-13795.

Cox, D.R. (1972). Regression models and life tables (with discussion). *Journal of the Royal Statistical Society Series B,* 34, 187-220.

Dudoit, S., Fridlyand, J., and Speed, T.P. (2002). Comparison of Discrimination Methods for the Classification of Tumors using Gene Expression Data. *Technical Report* #576. http://www.stat.berkeley.edu/~sandrine/tecrep/576.pdf.

Garber, M.E., Troyanskaya, O.G., Schluens, K., Peterson, S., Thaesler, Z., Pacyna-Gengelbach, M., van de Rijn, M., Rosen, G.D., Perou, C.M., Whyte, R.I., Altman, R.B., Brown, P.O., Botstein, D., Petersen, I. (2001). Diversity of gene expression in adenocarcinoma of the lung. *Proc. Natl. Acad. Sci. USA,* 98, 13784-13789.

Guyon, I., Weston, J., Barnhill, S., and Vapnik, V. (2002). Gene selection for cancer classification using support vector machines. *Machine Learning,* 46, 389-422.

Mateos, A., Herrero, J., Tamames, J., and Dopazo, J. Supervised and hierarchical unsupervised neural networks for clustering both gene expression profiles and samples. *Methods of Microarray Data Analysis II,* Lin, S., and Johnson, K., eds., Kluwer, Boston, 2001.

McLachlan, G.J., Bean, R.W., and Peel, D. (2002). A mixture model-based approach to the clustering of microarray expression data. *Bioinformatics,* 18, 413-422.

McLachlan, G.J. and Peel, D. (2000). Finite Mixture Models. Wiley, New York.

Wigle, D.A., Jurisica, I., Radulovich, N., Pintilie, M., Rossant, J., Liu, N., Lu, C., Woodgett, J., Seiden, I., Johnston, M., et al. (2002). Molecular profiling of non-small cell lung cancer and correlation with disease-free survival. *Cancer Research,* 62, 3005-3008.

This page intentionally left blank

Chapter 13

MICROARRAY DATA ANALYSIS OF SURVIVAL TIMES OF PATIENTS WITH LUNG ADENOCARCINOMAS USING ADC AND K-MEDIANS CLUSTERING

Wenting Zhou, Weichen Wu, Nathan Palmer, Emily Mower, Noah Daniels, Lenore Cowen, and Anselm Blumer
Computer Science, Tufts University, Medford, MA 02155 USA

Abstract: We experiment with two types of clustering, K-medians and a dimension-reduction technique known as approximate distance clustering (ADC) [Cowen and Priebe 1997], for classifying lung adenocarcinomas into high-risk and low-risk groups according to gene expression values from microarray data. The microarrays were Affymetrix oligonucleotide arrays used in studies at Michigan and Harvard, with 12,600 and 7129 probesets respectively. We show that we can obtain accurate classification based on a reduced set of genes obtained by nearest shrunken mean (NSM) [Tibshirani et al. 2002] or a combination of a variance-based approach with hierarchical clustering. The quality of the clustering is measured by using the p-values from log-rank tests, and the results are confirmed using cross-validation and by using the reduced set of genes obtained from one dataset to cluster the other.

Key words: Microarray; ADC clustering; K-medians; adenocarcinoma; survival time

1. INTRODUCTION

This paper investigates clustering and dimension-reduction techniques on two of the four CAMDA 2003 datasets of gene expression values and survival times of patients with lung adenocarcinomas. We chose the Michigan [Beer et al. 2002] and Harvard [Bhattacharjee et al. 2001] data due to the reasonably large sample sizes (n = 86 and 84) and lack of missing values. We use approximate distance clustering (ADC) maps [Cowen and

Priebe 1997] to project the data into one or two dimensions so we can use very simple clustering techniques, then follow this with nearest shrunken mean (NSM) [Tibshirani et al. 2002] to reduce the number of genes used to predict the clusters. We contrast this with more classical techniques of variance ratios and hierarchical clustering.

2. METHODS

2.1 Approximate Distance Clustering (ADC)

Approximate distance clustering (ADC) is a method that reduces the dimensionality of data by calculating the distances from data points to subsets of the data points called "witness sets" [Cowen and Priebe 1997]. One witness set is chosen for each desired output dimension.

It is defined as follows:

a) Let **X** be a collection of data in $\mathbf{R}^m$ In this case, each data point corresponds to a gene chip, so m is 12,600 or 7,129 initially.
b) Define $D_1, D_2, \ldots, D_d$ to be subsets of **X** of sizes $k_1, k_2, \ldots, k_d$. These are the witness sets.
c) The associated ADC map, $f_{(D1, D2, \ldots, Dd)}: \mathbf{R}^m \rightarrow \mathbf{R}^d$ maps **X** to $(y_1, y_2, \ldots, y_d)$, where $y_i = \min\{ \| x_j - x \| : x_j \varepsilon D_i \}$.

In other words, data point x maps to a point in m-dimensional space with i^{th} coordinate equal to the distance from x to the nearest point in the ith witness set. A good witness set is a small set of points that produces a mapping that preserves inter-cluster distances. In this paper, we look at the simplest cases of ADC projection on the microarray data: the case where the number of dimensions we project to is one or two, and the size of all witness set is one. Note that ADC does not in itself produce a clustering; the resulting points in one or two dimensions must still be classified or clustered using some method that works for low-dimensional data. In one dimension we just pick a cutoff value and assign all points below the cutoff to one cluster and all points above to the other. In two dimensions, we add the coordinates together before comparing to the cutoff. We use the following criterion to choose a good clustering from the set of allowable clusterings:

Compute the Kaplan-Meier survival curves and the p-value from the log-rank test, then use the following w-criterion:

$$w = 5500 * a + 4000 * b + 450 * (1-c) + 50 * d \qquad (1)$$

where

a is the p-value
b is 1 if the size of smaller group is less than $n/8$, and 0 otherwise
c is the difference between the final survival rates of the low-risk and high-risk groups
d is the high-risk group's final survival rate

This criterion is designed to select cluster separations with low p-values where neither cluster is too small and the final survival rates are reasonable. We use leave-one-out cross-validation to validate selecting the witnesses and cutoffs according to this criterion.

2.2 Nearest Shrunken Mean (NSM) Gene Reduction

After choosing the high-risk and low risk clusters using ADC clustering according to the w-criterion, we use nearest shrunken mean (NSM) [Tibshirani et al. 2002] to eliminate genes (or probesets) that have all their cluster means close to their overall mean.

Let:
x_{ij} be the expression of gene i for tissue sample j
C_k be class (or cluster) k
m_{ik} be the mean expression of gene i in C_k
x_i be the mean of gene i
n be the sample size
K be the number of clusters
n_k be the size of C_k
$s_i = (1/(n-K)) \Sigma_k \Sigma_{j\in Ck}(x_{ij}-m_{ik})^2$
s_0 be the median of the s_i
$M_k = \text{sqrt}(1/n_k+1/n)$
$d_{ik} = (m_{ik} - x_i) / (m_k{*}(s_i+s_0))$, so
$m_{ik} = x_i + d_{ik} \cdot m_k{*}(s_i+s_0)$

In this expression, d_{ik} can be reduced by Δ in absolute value or replaced by zero if its absolute value is smaller than Δ. If it is replaced by zero, the cluster mean becomes the overall mean; if this happens for all clusters, the gene can be eliminated.

2.3 K-medians Clustering

K-medians clustering is a variation of K-means clustering where the cluster centers must be chosen from among the data points. It is an unsupervised method, so the quality of the clustering is measured just using the distances between the data points without looking at their classifications. It selects K points to be cluster centers and calculates the quality of the clustering as the sum of the distances of data points to their nearest cluster center. In this paper, we use K=2 so it is feasible to calculate the quality of all $n(n+1/2$ clusterings and choose the optimal one.

2.4 Minimal Variance Ratio (MVR) Gene Reduction

The variance ratio is the sum of the within-cluster variances divided by the total variance of expression values for that gene. Using the notation from the NSM section above, let

a) $\sigma^2_{ik} = (1/n_k)\Sigma_{j\in Ck}(x_{ij}-m_{ik})^2$ be the within-cluster variance for gene i in cluster k.
b) $\sigma^2_i = (1/n)\Sigma_j(x_{ij}-x_i)^2$ be the total variance for gene i, then
c) $(\Sigma_k \sigma^2_{ik})/\sigma^2_i$ is the variance ratio for gene i.

Genes with large variance ratios are thought to contribute less to the cluster definitions and are eliminated.

2.5 Dimension Reduction With ADC and NSM

One set of experiments involved using one or two dimensional ADC clustering with a witness set of size one, followed by NSM to obtain a set of genes of the desired size. The w measure above was used to select the witness and the cutoff point between the two clusters. In the case of two dimensional ADC clustering we summed the values of the distances along the two axes to determine whether a point was below the cutoff. We also experimented with survival-time cutoff clustering (STCC), sorting the patients according to survival time and splitting them 50-50 or 60-40 into high risk – low risk clusters to replicate the results of [Beer et al. 2002].

2.6 Dimension Reduction With MVR, K-Medians, and Hierarchical Clustering

A second set of experiments involved starting with high-risk and low-risk clusters of equal size according to survival times (50% STCC), then using MVR to select a subset of genes to approximate this clustering. Some genes in this subset may have similar expression profiles, so a form of hierarchical clustering was used to obtain a desired number of clusters of these genes and one gene was selected from each cluster. This doubly reduced gene set was then used (after normalizing each gene profile to have vector length one) to obtain a K-medians clustering with K=2 and the p-value from the log-rank test was calculated.

3. EXPERIMENTAL RESULTS

We experimented with these methods on adenocarcinoma examples (patients) from the Michigan [Beer et al. 2002] and Harvard [Bhattacharjee et al. 2001] data that had survival times (both censored and uncensored). The Michigan data had expression values for 7,129 probesets for each of 86 examples, while the Harvard data had expression values for 12,600 probesets for each of 84 examples.

3.1 ADC on Harvard and Michigan data

Tables 1 through 4 give the results of using the w-criterion to select the best ADC witnesses and cutoffs, then reducing the set of probesets to the specified size with NSM. In all cases the witness sets had size one. The p-values were obtained from leave-one-out cross-validation on the reduced set of probesets. Specifically, ADC clusters were formed based on the reduced set of probesets, leaving out one patient, with the best ADC clustering being selected according to the w-criterion. The excluded patient was then classified as high-risk or low-risk according to which cluster mean was closer. The values for STCC were obtained by following the same procedure but substituting clusters formed of the 50% or 60% highest risk patients for the ADC clusters.

Table 1. p-values for one and two dimensional ADC and STCC on Michigan data (n = 86).

Genes	1D ADC	2D ADC	50% STCC	60% STCC
7129	0.0028	0.0500	0.0086	0.0126
1000	0.0275	0.0009	0.0111	0.0158
500	0.0495	0.0048	0.0046	0.0089
200	0.0019	0.0033	0.0075	0.0056
100	0.0058	0.0194	0.0023	0.0048
50	0.0019	0.1442	0.0064	0.0048
40	0.0009	0.0268	0.0011	0.0048
30	0.0009	0.0356	0.0029	0.0067
20	0.0021	0.0189	0.0029	0.0090
10	0.0061	0.0618	0.0059	0.0049
5	0.0086	0.3559	0.0151	0.0024

Table 2. Low risk/high risk group sizes for one and two dimensional ADC and STCC on Michigan data (n = 86).

Genes	1D ADC	2D ADC	50% STCC	60% STCC
7129	55/31	54/32	46/40	46/40
1000	59/27	60/26	45/41	43/43
500	52/34	57/29	47/39	45/41
200	58/28	58/28	47/39	48/38
100	57/29	55/31	49/37	46/40
50	58/28	42/44	50/36	47/39
40	58/28	44/42	50/36	47/39
30	58/28	43/43	51/35	46/40
20	57/29	42/44	51/35	46/40
10	56/30	37/49	50/36	47/39
5	58/28	41/45	49/37	49/47

Table 3. p-values for one and two dimensional ADC and STCC on Harvard data (n = 84).

Genes	1D ADC	2D ADC	50% STCC	60% STCC
12600	0.0646	0.0046	0.1946	0.0741
1000	0.0124	0.0013	0.0381	0.0038
500	0.0023	0.0116	0.0021	0.0027
200	0.0121	0.0037	0.0007	0.0004
100	0.0201	0.0027	0.0213	0.0004
50	0.0332	0.0090	0.0120	0.0047
40	0.0332	0.0019	0.0100	0.0033
30	0.0898	0.0010	0.0065	0.0098
20	0.0448	0.0039	0.0083	0.0015
10	0.0424	0.0011	0.0034	0.0001
5	0.0321	0.0032	0.0053	0.0196

Table 4. Low risk/high risk group sizes for one and two dimensional ADC and STCC on Harvard data (n = 84).

Genes	1D ADC	2D ADC	50% STCC	60% STCC
12600	25/59	24/60	39/45	41/43
1000	20/64	15/69	44/40	38/46
500	21/63	22/26	42/42	36/48
200	21/63	21/63	40/44	32/52
100	24/60	26/58	42/42	30/54
50	21/63	21/63	40/44	35/49
40	21/63	27/57	40/44	35/49
30	28/56	26/58	39/45	35/49
20	27/55	26/58	38/46	34/50
10	22/62	20/64	37/47	33/51
5	20/64	25/59	36/48	28/56

Since these datasets contained multiple probesets corresponding to the same genes, we then selected the top 50 probesets corresponding to distinct genes. Tables 5 and 6 give the probeset names, gene symbols, and mean expression values in the low-risk and high-risk group for each probeset selected. It is interesting to note that in the Michigan dataset most of these 50 (all except IGKC, IGL@, IGHG3, NPC2, HLA-A, CD74, HLA-B, MGP, NBL1, GRN, and the two with NULL symbol) have lower mean expression values in the low-risk group, while in the Harvard dataset all except GAPD, CLDN9, MIF, and PSMB3 have higher mean expression values in the low-risk group.

Cross-validation of the classification based on these expression values gave p-values of 0.0074 on the Michigan dataset and 0.0331 on the Harvard dataset. Figures 1 and 2 give the Kaplan-Meier curves corresponding to these p-values.

Table 5. Top 50 distinct genes from Michigan data. Underlined genes are also found in Table 6, bold genes are among the top 100 in [Beer et al. 2002].

Probeset	Symbol	Low-Risk	High-Risk
M63438_s_at	IGKC	29936.2	14461.4
M34516_at	NULL	23771.3	7285.7
X57809_s_at	IGL@	23693.4	6952.74
M87789_s_at	IGHG3	41259.8	8671.2
L19437_at	TALDO1	1352.48	2566.89
X01677_f_at	**GAPD**	8820.27	12018.6
L10678_at	PFN2	775.93	1462.43
X67698_at	NPC2	8877.69	6543.1
M21388_r_at	NULL	3370.06	2362.68
X00274_at	HLA-A	14115.9	11346.3
M13560_s_at	CD74	8951.48	6846.82
M17886_at	RPLP1	13417.8	19409.6
D49387_at	LTB4DH	372.44	1068.32
M37583_at	**H2AFZ**	1557.07	2302.42
X67951_at	PRDX1	4228.8	5964.1
X02152_at	LDHA	6607.16	8852.83
D13630_at	**KIAA0005**	1129.9	1655.69
D14874_at	**ADM**	368.88	624.67
X15940_at	RPL31	7048.57	8760.57
J03934_s_at	NQO1	481.3	1309.43
X91247_at	TXNRD1	1369.52	2603.73
X69654_at	**RPS26**	5012.86	6148.86
M22382_at	HSPD1	2687.07	3960.79
X77584_at	TXN	3019.61	4447.59
M26730_s_at	UQCRB	1783.05	2319.47
D49824_s_at	**HLA-B**	24959.3	18358.9
X15183_at	HSPCA	4756.56	6527.33
U09813_at	ATP5G3	2284.24	3336
X56468_at	YWHAQ	1832.02	2488.57
X13238_at	COX6C	1824.35	2530.02
D14657_at	KIAA0101	311.29	536.96
M22760_at	COX5A	1112.69	1458.31
D00762_at	PSMA3	1243.9	1629.8
J04823_rna1_at	COX8	4599.03	5722.32

Probeset	Symbol	Low-Risk	High-Risk
X53331_at	MGP	7151.91	4174.75
M24485_s_at	GSTP1	5788.36	8422.77
L08666_at	**VDAC2**	1480.34	2011.79
X65614_at	**S100P**	2495	6197.89
L37043_at	CSNK1E	858.41	1145.46
J04444_at	CYC1	1042.34	1524.23
M19961_at	COX5B	1631.52	2097.81
L19686_rna1_at	MIF	7390.13	8807.46
D28124_at	NBL1	4359.21	2358.11
X62320_at	GRN	3043.87	2825.88
Z14244_at	COX7B	461.46	705.04
Z49099_at	SMS	1017.55	1426.29
V00572_at	PGK1	3705.16	5137.71
U84573_at	PLOD2	555.49	710.12
U31814_at	HDAC2	421.74	611.64
HG4074-HT4344_at	FEN1	248.57	394.65

Table 6. Top 50 distinct genes from Harvard data. Underlined genes are also found in Table 5, bold genes are among the top 100 in [Beer et al. 2002].

Probeset	Symbol	Low-Risk	High-Risk
36627_at	SPARCL1	513.74	298.01
41723_s_at	**HLA-B**	1845.4	1001.59
38833_at	HLA-A	1936	1066.85
216_at	PTGDS	895.54	494.11
32905_s_at	TPSB2	454.99	193.21
39220_at	SCGB1A1	687.17	135.19
31525_s_at	HBA2	697.61	380.52
35905_s_at	**GAPD**	4541.9	5160.53
38691_s_at	SFTPC	4873	1276.4
32052_at	HBB	1032.3	580.48
32542_at	FHL1	121.61	52.95
1288_s_at	EEF1A1	5176.5	4636.86
35016_at	CD74	2641.5	1740.34
36097_at	ETR101	504.53	341.97
34363_at	SEPP1	322.25	182.02
1005_at	DUSP1	675.19	421.52
36634_at	BTG2	574.35	393.16
649_s_at	CXCR4	310.64	231.84
37394_at	C7	125.46	37.66
37021_at	CTSH	1988.2	1009.38

Probeset	Symbol	Low-Risk	High-Risk
33383_f_at	SFTPB	2232.6	1179.78
39864_at	CIRBP	353.61	276.04
35521_at	CLDN9	-91.05	14.98
31870_at	CD37	197.08	114.42
37168_at	LAMP3	304.36	84.97
41382_at	DMBT1	462.34	199
40607_at	DPYSL2	296.13	195.57
36495_at	FBP1	443.57	265.59
36669_at	FOSB	357.53	170.15
895_at	MIF	1270.2	1758.25
36680_at	AMY2B	242.38	56.31
534_s_at	FOLR1	782.5	449.16
36452_at	SYNPO	604.05	490.65
35183_at	ABCA3	376.66	152.54
428_s_at	B2M	3152.4	2805.04
39066_at	MFAP4	108.89	35.79
1915_s_at	FOS	1010	752.43
35926_s_at	LILRB1	1212.5	834.49
32321_at	HLA-E	481.98	365.18
34793_s_at	PLS3	321.7	217.19
35842_at	IL6ST	281.29	206.09
32786_at	JUNB	458.56	329
35730_at	ADH1B	43.05	15
31775_at	SFTPD	743.05	260.13
1117_at	CDA	312.02	209.11
1309_at	PSMB3	223.86	285.94
39345_at	NPC2	2083.5	1352.04
32597_at	RBL2	160.27	121.24
35868_at	AGER	139.54	54.72
33295_at	FY	124.02	79.66

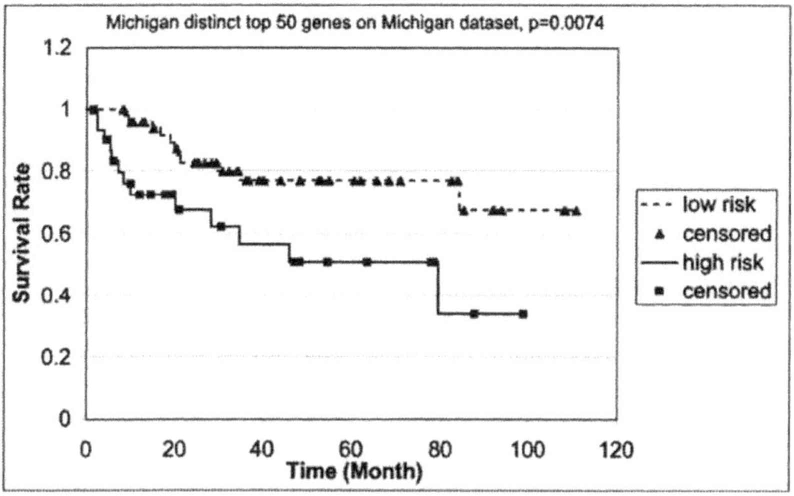

Figure 1. Kaplan-Meier curves for cross-validation of probesets corresponding to 50 distinct genes selected from Michigan dataset, validated on Michigan dataset. Low-risk and high-risk groups were separated using ADC and the w-criterion, then the top 50 distinct genes (according to NSM) were retained. See Table 5 for probeset names and gene symbols.

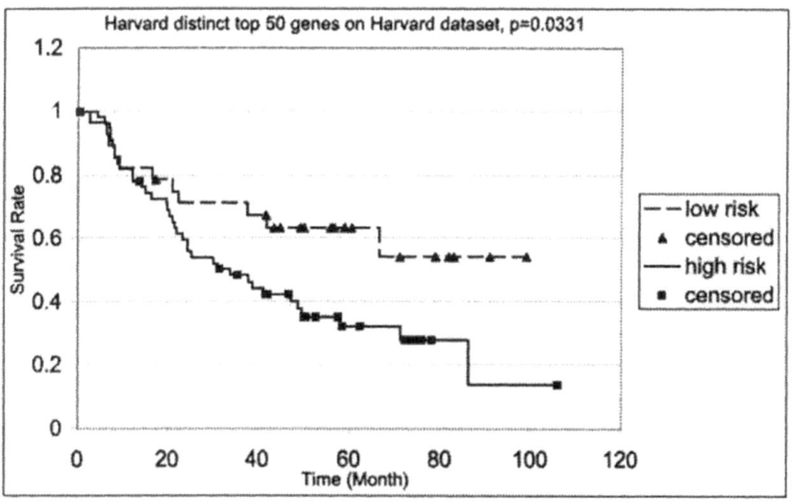

Figure 2. Kaplan-Meier curves for cross-validation of probesets corresponding to 50 distinct genes selected from Harvard dataset, validated on Harvard dataset. Low-risk and high-risk groups were separated using ADC and the w-criterion, then the top 50 distinct genes (according to NSM) were retained. See Table 6 for probeset names and gene symbols.

3.2 Validating ADC between Harvard and Michigan data

We also validated the groups of 50 probesets described above across datasets. Since the Michigan and Harvard studies used different gene chips, we used the probeset link table from affymetrix.com (filename PN600444HumanFLComp.zip) to find corresponding probesets in the two datasets. Starting from the top 50 probesets in the Michigan data we found the 57 matching probesets in the Harvard dataset, since the link table is not one-to-one. We then averaged probesets with the same gene symbol (including three with NULL symbol), leaving 48 distinct genes (plus NULL). We used those 49 as in the internal leave-one-out cross-validation to classify each example as low-risk or high-risk. Testing the top Michigan probesets on the Harvard data in this way gave a p-value of 0.0254. We then reversed this procedure, starting with the top 50 Harvard probesets. This gave 42 distinct genes in the Michigan dataset (plus NULL). Using those 43 for cross-validation on the Michigan data gave a p-value of 0.0307. Figures 3 and 4 give the Kaplan-Meier curves corresponding to these p-values.

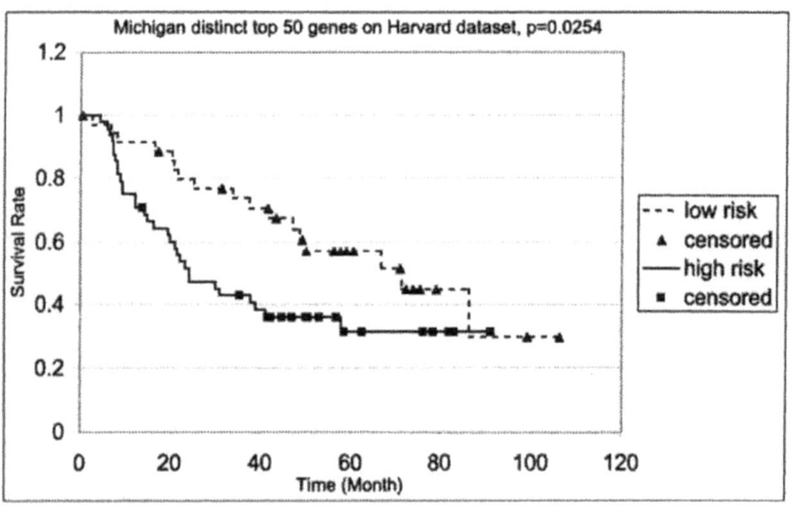

Figure 3. Kaplan-Meier curves for cross-validation of probesets corresponding to 50 distinct genes selected from Michigan dataset (see Table 5), validated on Harvard dataset by using equivalent probesets from the Harvard data.

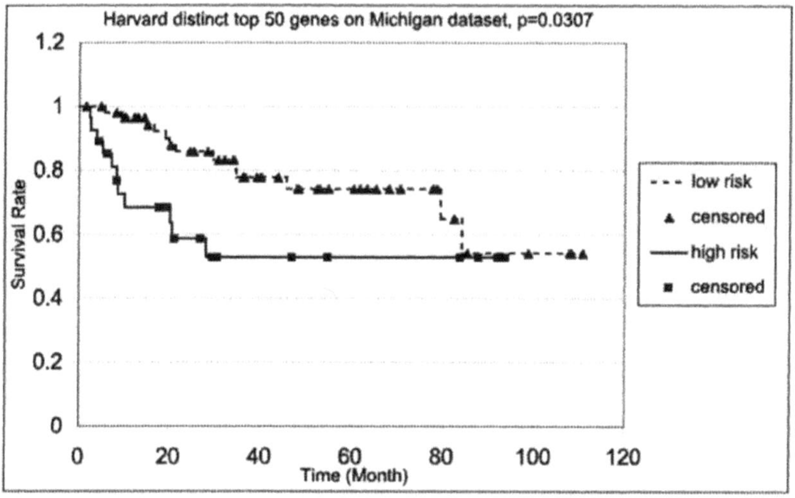

Figure 4. Kaplan-Meier curves for cross-validation of probesets corresponding to 50 distinct genes selected from Harvard dataset (see Table 6), validated on Michigan dataset by using equivalent probesets from the Michigan data.

3.3 MVR and K-medians

We used minimal variance ratio (MVR) to select 200 probesets from the Michigan and Harvard data based on an initial 50-50 clustering according to survival times (50% STCC), then used hierarchical clustering to group these probesets into 40 clusters. We selected one probeset from each cluster and performed a K-medians clustering of the patients into a high-risk and low-risk group using these 40 probesets after normalizing their expression profiles so that the clusters wouldn't be influenced unduly by probesets with high mean expression values. On the Michigan data this gave a p-value of 0.00002 with cluster sizes of 36 and 50, while on the Harvard data the p-value was 0.0417 with cluster sizes of 47 and 37. Kaplan-Meier curves for these are given in Figures 5 and 6.

We used leave-one-out cross-validation to verify this whole procedure. After clustering, the remaining patient was classified as high-risk or low-risk according to which cluster had the smaller average distance to that patient. For the Michigan data, this gave a p-value of 0.0219 and for the Harvard data the p-value was 0.0696.

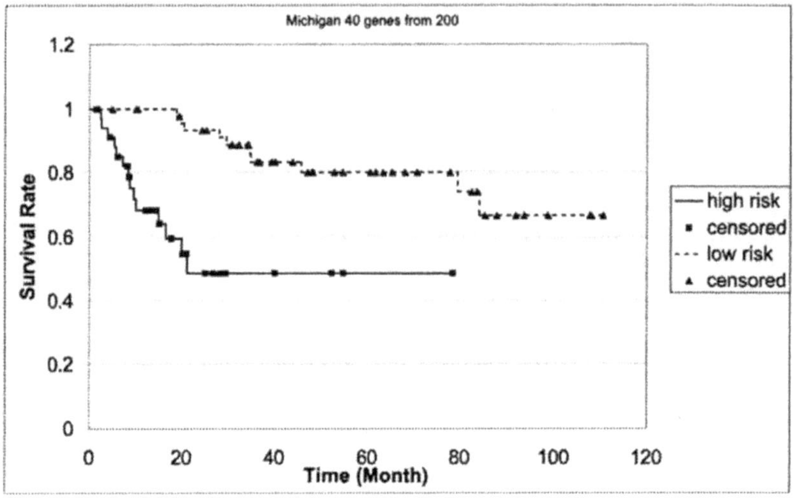

Figure 5. Kaplan-Meier curve for classifying Michigan data according to 40 probesets selected using MVR, K-medians, and hierarchical clustering of probesets.

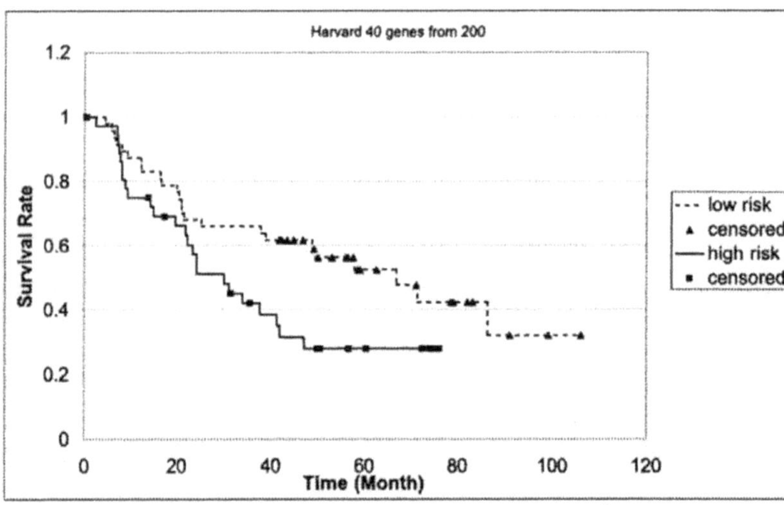

Figure 6. Kaplan-Meier curve for classifying Harvard data according to 40 probesets selected using MVR, K-medians, and hierarchical clustering of probesets.

4. CONCLUSIONS

On the Michigan data one-dimensional ADC clustering obtained results very comparable in terms of the p-values of the Kaplan-Meier curves to those obtained by Beer using Cox model regression, and we were able to reduce the set of genes further than they reported [Beer et al. 2002]. Beer reported a p-value of 0.0006 for leave-one-out cross-validation based on a set of 50 genes, whereas in Table 1 we show p-values of 0.0009 for sets of 30 or 40 genes. On the Harvard data we obtained good results using two dimensional ADC, as reported in Table 3. We also obtained reasonable cross-validation between the Harvard and Michigan data.

Our reduced sets of genes differed significantly from those reported by Beer et al. [2002]. This is perhaps not surprising since our MVR and K-median experiments found that hierarchical clustering of the genes could often significantly reduce the number of genes without much of a decrease in the quality of the clustering as measured by the p-value. This probably indicates that the data contained many genes with closely related biological function. The following genes that have been associated to cancer appear on one or both of our top 50 lists, but were not among the top 50 reported by Beer:

a) SPARCL1 (also known as MAST9 or hevin) - down regulation of SPARCL1 also occurs in prostate and colon carcinomas, suggesting that SPARCL1 inactivation is a common event not only in NSCLCs but also in other tumors of epithelial origin. (http://www.ncbi.nlm.nih.gov:80/entrez/query.fcgi?cmd=Retrieve&db=PubMed&list_uids=11179481&dopt=Abstract)

b) CD74 - well-known for expression in cancers (http://biz.yahoo.com/prnews/031120/nyth078_1.html)

c) PRDX1 - linked to tumor prevention (http://www.ncbi.nlm.nih.gov:80/entrez/query.fcgi?cmd=Retrieve&db=PubMed&list_uids=12891360&dopt=Abstract)

d) PFN2 - seen as increasing in gastric cancer tissues (http://cancerres.aacrjournals.org/cgi/content/full/62/1/233)

e) SFTPC - responsible for morphology of the lung; a mutation causes chronic lung disease (http://www.ncbi.nlm.nih.gov:80/entrez/query.fcgi?cmd=Retrieve&db=PubMed&list_uids=14525980&dopt=Abstract)

f) HLA-DRA (HLA-A) - lack of expression causes cancers (http://www.ncbi.nlm.nih.gov:80/entrez/query.fcgi?cmd=Retrieve&db=PubMed&list_uids=12756506&dopt=Abstract)

Not much is known about the function of the following genes: PTGDS, H2AFZ, KIAA0005 (also called BZW1), EEF1A1, TXNRD1, RPS26. The fact that appeared on our lists indicates that they may be worth further investigation.

These techniques provide computationally efficient ways to reduce a large set of genes or probesets to find ones of potential biological interest or to apply further statistical techniques that would be computationally infeasible on the larger dataset. We have suggested a couple of combinations, but others (such as ADC and NSM followed by hierarchical clustering) are also potentially useful and should be investigated further.

Source code for our programs (in C++) and further results are available from http://camda.cs.tufts.edu

5. REFERENCES

Beer, DG, Kardia SL, Huang CC, Giordano TJ, Levin AM, Misek DE, Lin L, Chen G, Gharib TG, Thomas DG, Lizyness ML, Kuick R, Hayasaka S, Taylor JM, Iannettoni MD, Orringer MB, Hanash S. Gene-expression profiles predict survival of patients with lung adenocarcinoma, 2002, *Nature Medicine* **8**(8):816-824.

Bhattacharjee, A., Richards, WG, Staunton J, Li C, Monti S, Vasa P, Ladd C, Beheshti J, Bueno R, Gillette M, Loda M, Weber G, Mark EJ, Lander ES, Wong W, Johnson BE, Golub TR, Sugarbaker DJ, and Meyerson M. 2001, Classification of human lung carcinomas by mRNA expression profiling reveals distinct adenocarcinoma subclasses, *PNAS* **98**(24):13790-13795.

Cowen, L. J. and Priebe, C.E., 1997, Randomized non-linear projections uncover high-dimensional structure. *Adv. Appl. Math.*, **19**:319-331.

Tibshirani, R., Hastie, T., Narasimhan, B., and Chu, G., 2002, Diagnosis of multiple cancer types by shrunken centroids of gene expression., *PNAS* **99**(10):6567-657.

Chapter 14

HIGHER DIMENSIONAL APPROACH FOR CLASSIFICATION OF LUNG CANCER MICROARRAY DATA

F. Crimins, R. Dimitri, T. Klein, N. Palmer and L. Cowen
Department of Computer Science, Tufts University

Abstract: A lung cancer microarray dataset is re-examined using simple techniques, but retaining more of the high-dimensional structure. In particular, instead of discarding genes that look uninformative when considered in isolation, pairs, triples and quartets of genes are selected using kNN classifiers. Genes of potential biological importance are also uncovered.

Key words: Lung cancer, microarray, classification, high-dimensional data

1. INTRODUCTION

The CAMDA 2003 competition involved the re-analysis of lung cancer microarray datasets. Four separate studies sought clusters that correlated with survival among patients diagnosed with adenocarcinoma, the most common type of lung cancer. We shall consider here two of the datasets, those of Bhattacharjee et al. [2001] and Garber et al. [2001], that included samples not only from adenocarcinoma, but also microarray data from several other types of lung cancer tumors, as well as normal lung tissue. A secondary goal was to construct a classifier that could distinguish between expression data for each of several different lung-cancer types plus normal lung tissue, and, in addition, find a small set of expression vectors that could account for the difference. It is this easier classification problem that is the subject of the present paper, and we reanalyze the results of both Bhattacharjee et al. [2001] and Garber et al. [2001]. A second paper by Zhou et al. [2004], also published in this volume, represents our approach to the more complicated question of predicting survival outcomes.

Bhattacharjee et al. [2001] obtained a dataset of 186 patients with four clinically distinct types of lung cancer plus normal lung tissue. The data consisted of expression values for 12,600 transcript sequences for each patient. The dataset of Garber et al. [2001] consisted of expression levels for 73 samples over 23,100 transcript sequences (representing 17,108 unique genes). Of these samples, 67 were taken from tumors of four different cancer types, five were from normal tissue and one was taken from fetal lung tissue. We have included all of the normal tissue samples, and excluded the fetal lung tissue sample. Eleven of the tumors were sampled twice. In these instances, we omitted the samples taken from peripheral biopsies, intrapulmonary metastases, and metastatic lymph nodes in favor of those taken from central biopsies and primary tumors. One of these 11 pairs consisted of two intrapulmonary metastases from the same patient, who was also represented by a primary tumor sample; in keeping with our practice of examining only primary tumor samples, both of these intrapulmonary metastases were omitted in favor of the primary tumor sample. Finally, one of the remaining 56 tumor samples was diagnosed as combined LCLC and SCLC; this sample was omitted because there was no clear rule for determining correct class prediction. This resulted in a dataset consisting of expression information for 59 samples: 34 AD, four LCLC, five Normal, 12 SCC and four SCLC (three of which were equivalent to three of the four considered by Bhattacharjee et al. [2001]).

Because the number of transcript sequences was very large, both groups of researchers first identified a subset of transcript sequences that had "meaningful" expression data. In particular, Bhattacharjee et al. [2001] identified a subset of 3,312 "most variable" genes over the five different lung tissue types, and then used a subset of 675 of these to construct subclusters of the adenocarcinoma subtype. Garber et al. [2001] searched transcript sequences that were similar among tumor pairs, but varied most widely among all tumor samples, yielding a subset of 918 (representing 835 unique genes).

In this paper, we show that there is something to be gained by studying the entire dataset without first doing such preliminary dimension reduction—that such an approach can yield better results than immediately restricting to a small subset of transcript vectors whose values, considered individually, appear to contain the most discriminatory information. In this sense, our work supports the study of Li et al. [2001] in CAMDA 2000, which also cautioned against looking at single expression vectors in isolation, in order to determine their discriminatory power.

While processing 12,600-dimensional data is beyond the scope of most commercial software packages designed for these problems, it is only a minor headache to code simple non-parametric classifiers that can handle the full dimensionality of the dataset in C++. First we show that perhaps the simplest non-parametric classifier, k-nearest-neighbor, performs extremely well on the 5-class problem considered by Bhattacharjee et al. [2001] in cross-validation. That this is such an easy problem is perhaps not too surprising, given that these classes of lung tissue are also clinically identifiable as distinct. We then go on to the more interesting question of finding a small set of transcript sequences whose expression profiles can, by themselves, discriminate among the five classes. This is important because small sets of transcript sequences that distinguish among the classes may correspond to genes that are biologically important toward understanding the underlying lung-cancer pathology. Combinatorial complexity quickly prohibits an exhaustive search of all r-element subsets of transcript sequence responses, for even small values of r. We show, however, that by moving from r=1 to r=2, and considering pairs of expression vectors in concert, we achieve interesting results and identify seemingly biologically important genes that were not identified by previous analyses. When the approach is bootstrapped to r=3 and then r=4, we find, for the dataset of Bhattacharjee et al. [2001] nine 4-tuples of transcript sequences that each yield 97% correct classification for the 5-class problem. We show that the same method applied to the dataset of Garber et al. [2001] yields five 4-tuples of transcript sequences that each yield at least 97% correct classification among the six classes of lung cancer tumor contained in that dataset. Again, these results can be seen as validating the method of Li et al. [2001] presented at CAMDA 2000; Li et al. [2001] use a genetic algorithm to heuristically explore the space of r-element subsets. For the data dimensionality and the small value of r considered here, we were able to find the best subsets exactly, using exhaustive search. However, for larger datasets or larger r, a genetic algorithm or other heuristic search approach such as Li et al. [2001] use, becomes appropriate.

1.1 k-Nearest Neighbor Classifiers

The k-nearest-neighbor classifier, first introduced by Fix and Hodges [1951], is the simplest and best-known non-parametric classifier. It is based on a distance, or dissimilarity measure d, that is assigned to all pairs of observations; for this study we used the L1 metric, where if $\mathbf{x} = (x_1, \ldots, x_n)$, and $y = (y_1, \ldots, y_n)$ are observations, then $d(x,y) = \Sigma \mid x_i - y_i \mid$.

The kNN classifier is typically defined as follows. Suppose training data $\mathbf{T} = \{t_1, \ldots, t_r\}$ are a set of observations labeled by their class labels from $C = \{c_1, \ldots, c_m\}$. Let x be an observation whose class label is unknown. Define $\mathbf{S}_x \subseteq \mathbf{T}$ to be the set of x's k closest neighbors according to the distance metric d in T. Assuming no ties, let c_i be the class label that appears most frequently in the set $\mathbf{S_x}$. Then x is assigned the class label c_i (notice in the case that there are two classes and k is odd, there will be no ties). In the case that more than one class label appears with equal frequency in $\mathbf{S_x}$, we regress to the (k-1)NN classifier; if there is again a tie we regress to the (k-2)NN classifier, and so on. This must result in a unique class name, no matter how many classes there are, since when k=1 there can be no ties.

Given the raw 12,600-dimensional dataset provided by Bhattacharjee et al. [2001], patients were divided into five groups, with each patient assigned to the group corresponding to his index mod 5. (Since the patients were grouped by class in the data set this was not a random partition, but rather had the effect of spreading out the number of patients of each class in each group as close to evenly as possible). We first show that, without any re-normalizing, pre-processing, or scaling, the 5-nearest-neighbor classifier correctly classifies 94% of the patients by lung tissue type in a 5-fold cross-validation (see Table 1). From this we conclude that the 5-class problem of separating tissue samples into adenocarcinomas, squamous, SCLC, pulmonary carcinoid, and normal lung, is at most a problem of moderate difficulty. This is not surprising, given that the different classes are considered clinically distinct [Bhattacharjee et al., 2001],

Table 1. kNN 5-fold cross-validation on the entire 12,600-dimensional data set.

	1 kNN % correct	3 kNN % correct	5 kNN % correct	7 kNN % correct
Group 1	95.1219	92.6829	95.1219	92.6829
Group 2	85.3659	87.8049	85.3659	85.3659
Group 3	90.2439	90.2439	92.6829	90.2439
Group 4	90.0000	97.5000	97.5000	90.0000
Group 5	95.0000	97.5000	100.0000	92.5000
Average	91.1330	93.5961	94.0887	90.1478

2. TWO CLASS SUB-PROBLEMS

We can also ask about the best genes for distinguishing each of the five individual classes from their complements. These problems vary in difficulty. In particular, there are six different transcript sequences that individually separate the pulmonary carcinoids from the other classes with 100% accuracy, and another three that achieve 99.5% accuracy, using 1NN in a leave-one-out cross-validation. Similarly, the gene Hs.181163 – described as high-mobility group (non-histone chromosomal) protein 17, gives complete separation with 1NN between the SCLC class and all the rest, while an additional five transcript sequences give > 99.5% correct classification. The best individual sequence to separate the normal samples from all tumorous classes achieves 99% classification using the same method, while the top ten sequences all achieve > 97% classification. For the squamous class, there is one individual transcript sequence that achieves 98% correct classification (it is gene Hs. 137569, tumor protein 64 kDa with strong similarity to p53, previously known to be a signature for squamous tumors [Bhattacharjee et al., 2001]), and then there is a gap in discriminatory power. However, the top ten individual transcript sequences all achieve > 93% classification. With respect to the adenocarcinoma class, the best individual transcript sequence still achieves slightly less than 81% correct classification when separating the adenocarcinomas from the other four classes using 5-nearest-neighbor.

A list of the top individual genes and their corresponding gene names for classifying each of these tumor types appears below.

Table 2. The best individual transcript sequences to indicate membership/nonmembership in the pulmonary carcinoid class.

UNIGENE ID	Description	% Correct
Hs.124411	chromogranin A (parathyroid secretory protein 1)	100.0%
Hs.172740	microtubule-associated protein, RP/EB family, member 3	100.0%
Hs.25348	Cluster Incl AL050223:Homo sapiens mRNA; cDNA DKFZp586L1323	100.0%
Hs.74565	Cluster Incl U48437:Human amyloid precursor-like protein 1 mRNA	100.0%
Hs.136164	cutaneous T-cell lymphoma-associated tumor antigen se20-4	100.0%
Hs.89655	protein tyrosine phosphatase, receptor type, N	100.0%
Hs.323833	syntaphilin	99.5%
Hs.304330	KIAA0656 gene product	99.5%
Hs.148258	KIAA0430 gene product	99.5%

Table 3. The best individual transcript sequences to indicate membership/nonmembership in the SCLC class.

UNIGENE ID	Description	% Correct
Hs.181163	high-mobility group (nonhistone chromosomal) protein 17	100.0%
Hs.443960	Cluster Incl U75968:Human clone C3 CHL1 protein (CHLR1) mRNA	99.5%
Hs.505	ISL1 transcription factor, LIM/homeodomain, (islet-1) †	99.5%
Hs.77204	centromere protein F (350/400kD, mitosin)	99.5%
Hs.28853	CDC7 (cell division cycle 7, S. cerevisiae, homolog)-like 1	99.5%
Hs.396393	ubiquitin carrier protein	99.5%

Table 4. The best individual transcript sequences to indicate membership/nonmembership in the normal lung sample class. † indicates that the transcript sequence was previously identified as biologically important in Garber et al. [2001]; ‡ that it was identified in Bhattacharjee et al. [2001]. A probe set identifier in brackets indicates that no UNIGENE ID was available for the probe set.

UNIGENE ID	Description	% Correct
Hs.78146	Cluster Incl AA100961:zn40b06.s1 Homo sapiens cDNA	99.0%
Hs.65424	36569 at tetranectin (plasminogen-binding protein) ‡	98.5%
Hs.184	advanced glycosylation end product-specific receptor ‡	98.0%
Hs.511911	epithelial membrane protein 2	98.0%
Hs.76206	cadherin 5, type 2, VE-cadherin (vascular epithelium)	97.5%
Hs.421383	four and a half LIM domains 1	97.5%

UNIGENE ID	Description	% Correct
[1814_at]	transforming growth factor, beta receptor II (70-80kD) †	97.5%
Hs.155106	receptor (calcitonin) activity modifying protein 2	97.0%
Hs.333383	ficolin (collagen/fibrinogen domain-containing) 3 (Hakata antigen)	97.0%
Hs.89640	TEK tyrosine kinase, endothelial	97.0%

Table 5. The best individual transcript sequences to indicate membership/nonmembership in the squamous class. † indicates that the transcript sequence was previously identified as biologically important in Garber et al.[2001]; ‡ that it was identified in Bhatterjee et al. [2001]. A probe set identifier in brackets indicates that no UNIGENE ID was available for the probe set.

UNIGENE ID	Description	% Correct
Hs.137569	tumor protein 63 kDa with strong homology to p53 †‡	98.0%
Hs.349499	Cluster Incl AL031058:Human DNA sequence	94.5%
Hs.501990	Cluster Incl AA010777:ze22f06.r1 Homo sapiens cDNA, 5 end	93.5%
Hs.412999	Cluster Incl AA570193:nf38c11.s1 Homo sapiens cDNA	93.1%
Hs.443518	bullous pemphigoid antigen 1 (230/240kD) †	93.1%
Hs.355827	novel putative protein sim. to YIL091C yeast hyp. 84 kD prot.	93.1%
Hs.291385	Cluster Incl AF035315:Homo sapiens clone 23664 and 23905 mRNA seq.	93.1%
Hs.82237	ataxia-telangiectasia group D-associated protein †	93.1%
Hs.55279	serine (or cysteine)	93.1%

UNIGENE ID	Description	% Correct
	proteinase inhibitor, clade B (ovalbumin), member 5	
[601_s_at]	keratin 16 (focal non-epidermolytic palmoplantar keratoderma)	93.1%

Table 6. The best individual transcript sequences to indicate membership/nonmembership in the adenocarcinoma class. † indicates that the transcript sequence was previously identified as biologically important in Garber et al. [2001]; that it was identified in Bhatterjee et al. [2001]. A probe set identifier in brackets indicates that no UNIGENE ID was available for the probe set.

UNIGENE ID	Description	% Correct
Hs.446352	v-erb-b2 avian erythroblastic leukemia viral oncogene homolog 2	80.8%
Hs.75243	bromodomain-containing 2	80.3%
Hs.129729	ligand of neuronal nitric oxide synthase w/carboxyl-terminal PDZ domain	79.3%
[1814_at]	transforming growth factor, beta receptor II (70-80kD) †	78.8%
Hs.3192	Cluster Incl AA631698:np79a08.s1 Homo sapiens cDNA	78.3%
Hs.446352	v-erb-b2 avian erythroblastic leukemia viral oncogene homolog 2	78.3%
Hs.386467	intercellular adhesion molecule 1 (CD54), human rhinovirus receptor †	77.3%
Hs.446375	microtubule-associated protein, RP/EB family, member 2	77.3%
Hs.87417	cathepsin L2	77.3%

3. IDENTIFYING GENES THAT JOINTLY DISCRIMINATE

Constructing a good classifier is not the only problem one wants to solve with microarray data, particularly for this problem where the different classes are clinically distinct. A more interesting biological problem is to identify a small subset of genes whose expression signatures themselves distinguish among different classes. Then these genes can be hypothesized to be involved biologically in the cancer pathology of the cell. We discuss this problem now.

As discussed in Section 1, both prior studies on the multiple classes of lung cancer [Bhattacharjee et al., 2001; Garber et al., 2001] began by pre-filtering individual transcript sequences to come up with a subset of relevant genes. In contrast, Li et al. [2001], suggested that additional power in feature selection for microarray data can be obtained by considering small subsets of transcript sequences that jointly discriminate. This approach has demonstrably more power; however the computational obstacles grow quickly as the size of these subsets grow. In particular, to examine all pairs of transcript sequences in the 12,600-dimensional array requires 79,373,700 significance calculations, while to examine all triples of transcript sequences requires 333,316,624,200 significance calculations. For this reason, Li et al. [2001] suggest a genetic algorithms approach to intelligently search this space, and give results and sensitivity of their methods to initial starting conditions in Li et al. [2001].

We take a more elementary bootstrapping approach to identify these joint discrimination sets as follow. First we examine all unique pairs of transcript sequences in the dataset, and retain the 1024 best pairs. Then those 1024 best pairs are matched with all unique third transcript sequences in the dataset, and the best 512 triples are maintained. Finally, the strongest 512 triples are matched with all unique fourth transcript sequences to obtain the best 4-dimensional classifier.

In the above description, "best" must be determined based on some measure of discriminatory power. We employ k-nearest-neighbor again, and look at the percentage of correct classification based on a 1-nearest-neighbor classifier in a leave-one-out cross-validation to determine the quality of each low-dimensional projection. We first tried the method on the dataset of Bhattacharjee et al. [2001]. As we raised the dimensionality of the model, the classification rate improved. The best of the transcript sequence pairs was capable of classifying 89% (182/203) of the observations correctly in a leave-one-out cross-validation. Of the 3-dimensional classifiers examined, six were found capable of correctly classifying 94% of the observations. Finally, of the 4-dimensional models examined, nine 4-tuples were found that were capable of correctly classifying 97% (197/203) or more of the observations. Due to computational issues, we did not test all triples or 4-tuples: rather, the subset of triples that we tested was based on the best performing pairs, and the subset of 4-tuples we tested was based on the best performing triples (in a leave-one-out cross-validation). That is, while the triples chosen were based on an analysis of all possible gene pairs, the 4-tuples examined were based on selected triples. There is, therefore, a possible issue of selection bias aiding us in locating these best 4-tuples so quickly, as they were chosen based on the triples that performed the best over the entire dataset.

The list of the twelve most frequent genes occurring in the set of the 512 strongest triples appears in Table 7. The set of the nine best 4-tuples appears in Table 8. Information about the biological significance of some of these genes appears in Section 4. A longer version of Table 7 and more biological information can be found at http://www.cs.tufts.edu/~cowen/camda.

Table 7. Frequently occurring transcript sequences among top triples, with their frequency in the top triples and pairs. A probe set identifier in brackets indicates that no UNIGENE ID was available for the probe set.

UNIGENE ID	Frequency in top 512 triples	Frequency in top 1024 pairs
[1814_at]	273	197
Hs.24040	108	161
Hs.137569	59	10
Hs.74624	48	18
Hs.74565	37	7
Hs.78146	27	23
Hs.154658	26	24
Hs.75667	20	4
Hs.25640	19	13
Hs.349499	16	19
Hs.7979	16	12
Hs.184	15	28

(We remark that two of the best 4-tuples do very poorly on the SCLC class; whereas we showed that distinguishing SCLC observations from the others was among the easiest of the 2-class problems. Thus, by separately classifying non-SCLC and SCLC observations first, we could improve the classification rate to 98.5% (200/203). Alternately, combining one of these best 4-tuples (the first one in Table 8) with the single gene Hs.505 (a good classifier for SCLC), results in a 5-dimensional subset for which 1-nearest-neighbor classifies 98% (199/203) correctly.)

Table 8. List of the nine best 4-dimensional transcript sequence classifiers. A probe set identifier in brackets indicates that no UNIGENE ID was available for the probe set.

Classifier	AD (139)	NL (17)	SCLC (6)	SQ (21)	COID (20)	Total
Hs.20447, [1814_at], Hs.389, Hs.367725	138	17	3	20	20	97.5%
Hs.77204, Hs.24040, Hs.137569, Hs.151413	136	16	6	19	20	97%
Hs.137569, Hs.24040, Hs.75061,Hs.449098	137	17	3	20	20	97%
Hs.137569, Hs.24040, Hs.300684, Hs.446352	137	16	6	18	20	97%
Hs.74565, Hs.418123, Hs.446352, Hs.287850	137	16	4	20	20	97%
Hs.77204, Hs.24040, Hs.137569, Hs.2171	137	16	6	18	20	97%
[1814_at], Hs.292511, Hs.437508, Hs.505	137	16	6	18	20	97%
Hs.77204, Hs.24040, Hs.137569, Hs.193725	136	16	6	19	20	97%
Hs.137569, Hs.24040, Hs.9754, Hs.436301	136	16	6	19	20	97%

The nine best 4-tuples in Table 8 contain within them 22 unique transcript sequences; of these Hs. 137569, Hs.24040, Hs.446352, Hs.77204, and probe set 1814_at occur multiple times across the nine 4-tuples (in fact, Hs.137569 and Hs.24040 occur as a pair in six of the nine best 4-tuples). This argues for the biological importance of these genes in lung-cancer, particularly those that occur multiple times on the list, and in fact Hs.137569 is tumor protein 63 kDa with strong similarity to p53, involved in cell growth regulation, known to be involved in lung cancer pathology, and previously identified as critical by both Bhattacharjee et al. [2001] and Garber et al. [2001] . On the other hand, based on these results, we conjecture that Hs.24040, identified as potassium channel, subfamily K, member 3, that encodes one of the superfamily of potassium channel proteins [Duprat et al., 1997] is also of biological importance, and this gene was not flagged in any previous study. The biological meanings of the other genes that make up the classifiers in Table 8 is discussed briefly in Section 4, more details can be found at http://www.cs.tufts.edu/~cowen/camda.

The results in Table 8 represent the best 4-tuples selected by leave-one-out cross validation on the entire dataset. To address the issue of selection bias, we re-ran this expeiment, partitioning the data evenly into a training and a test set. The results were quite stable: the top two 4-tuples located in this analysis, both of which achieved 100% correct classification on the training data, achieved 94% and 93% correct classification on the test set, and included the genes Hs.74565 and Hs.24040. In fact, of the 1520 top 4-tuples that scored above 98% on the training data, 920 contained genes previously identified in Table 8.

To validate the method, we turned to the second data set of Garber et al. [2001]. We show that our method has similar performance on the 6-class problem of dataset 2. Table 9 shows the performance of the top five 4-dimensional classifiers by gene accession number, each of which correctly classifies 58 of the 59 patients, greater than 98% correct classification rate on the 5-class problem considered.

Once again, we suggest that the genes that show up multiple times in this table have biological significance for lung cancer pathology. In particular, we discuss what is known about R70462, H65075 and T84152 in Section 4.

Table 9. List of the five best 4-dimensional transcript sequence classifiers.

Classifier	AD (34)	LCLC (4)	NL (5)	SCC (12)	SCLC (4)	Total
R70462, H97677, R26186, AA007308	34	3	5	12	4	98.3%
R70462, AA862435, H65065, T84152	34	4	5	11	4	98.3%
R70462, T47454, N55459, AA460571	34	3	5	12	4	98.3%
R70462, H02848, H65065, H77706	34	3	5	12	4	98.3%
R70462, AA186348, H6505, T84152	33	4	5	12	4	98.3%

4. BIOLOGICALLY SIGNIFICANT GENES

We suggest that the transcript sequences that occur most frequently in Tables 7, 8, and 9, are biologically significant in lung cancer pathology. A full description of what is known about these genes for both datasets appears in supplementary information at http://www.cs.tufts.edu/~cowen/camda.

Two of the genes we find are also explicitly identified not only by our methods, but also in the papers of Bhattacharjee et al. [2001] and Garber et al. [2001]. These are probe set 1814 (transforming growth factor, beta receptor II), and gene Hs. 137569 (tumor protein 63 kDa with strong homology to p53). Both are known to be involved in the pathology of multiple cancers [Hibli et al., 2000; Markowitz et al., 1995].

We additionally find the gene Hs.446352, which occurs in one of the top 4-tuples for the dataset of Bhattacharjee et al. [2001], and is the same as R70462, which occurs in all the top triples in the dataset of Garber et al.

[2001]. The paper of Bhattacharjee et al. [2001] does not list this as an important gene at all; in the paper of Garber et al. [2001] it is on a list of nearly 500 genes that they identify as having a high expression value in all adenocarcinomas, but a low expression value in all squamous samples. The gene is v-erb-b2 erythroblastic leukemia viral oncogene homolog 2, neuro/glioblastoma derived oncogene homolog (avian). It encodes a tumor antigen, p185, which is serologically related to EFGR, the epidermal growth factor receptor [Yang-Feng et al., 1985]. Its role in cancer has been studied in several other papers, for example, van de Vijver et al. [1988] found that its over-expression in cancers corresponds to poor prognosis, enhanced metastatic potential, and chemoresistance.

Most of the frequently occurring genes in Tables 7, 8, and 9 are not identified as important in either of the papers of Bhattacharjee et al. [2001] and Garber et al. [2001], since they were excluded during the preprocessing done in each analysis. However, T84152, caveolin 2, has recently been implicated to have some role in cancer in the biology literature. Fong et al. [2003] show a positive correlation of the expression of caveolin 1 and caveolin 2 with tumor grade and squamous features of urothelial carcinoma. They suggest that caveolin 1 and caveolin 2 be studied further to determine a possible role in tumor progression and squamous differentiation. Other genes that appear important in our analysis but have not been previously identified as such by Bhattacharjee et al. [2001] and Garber et al. [2001] are Hs.24040, identified as potassium channel, subfamily K, member 3, that encodes one of the superfamily of potassium channel proteins [Duprat et al., 1997] and H65065, visinin-like 1, also referred to as VILIP1, which Lin et al. [2002] show modulates the surface expression and agonist sensitivity of the alpha 4 beta 2 nicotonic acetylcholine receptor in response to changes in levels of calcium. Minna [2003] links the alpha 4 beta 2 acetylcholinic receptors to lung cancer directly, claiming that smoking addiction is a result of the action of nicotine on these receptors.

5. CONCLUSIONS

We have shown that the simplest non-parametric classifiers can have some utility for some microarray classification problems, acting on the entire non-dimension reduced dataset. For the problem of determining small sets of transcript sequences that have discriminatory power (and thus possible significance in the biological pathway), we show that increasing the dimensionality of these sets (considering pairs, triples or 4-tuples, rather than individual transcript sequences one by one) can lead to significant improvements with each dimension gained. As a result, we caution the practitioner against reducing the dimensionality of the data too quickly.

ACKNOWLEDGEMENTS

Thanks to Donna Slonim for her encouragement and for reading an early draft of this paper.

REFERENCES

A. Bhattacharjee, W. Richards, J. Shaunton, C. Li, S. Monti, P. Vasa, C. Ladd, J. Beheshti, R. Bueno, M. Fillette, M. Loda, G. Weber, E. Mark, E. Lander, W. Wong, B. Johnson, T. Golub, D. Sugarbaker, and M. Meyerson (2001) Classification of human lung carcinomas by mRNA expression proliling reveals distinct adenocarcinoma sublcasses. PNAS, 98, 24, 13790-13795.

F. Duprat, F. Lesage, M. Fink, R. Reyes, C. Heurtenaux, and M. Lazdunski (1997) TASK, a human background K+ channel to sense external pH variations near physiological pH, Ebo J., 16, 5464-6471.

E. Fix and J. Hodges (1951) Discriminatory analysis: Nonparametric discrimination: Consistency properties. Technical report 21-49-004, USAF School of Aviation Medicine.

A. Fong, E. Garcia, L. Gwynn, M. Lisanti, M. Fazzari, and N. Li (2003) Expression of Caveolin-1 and Caveolin-2 in urothelial carcinoma of the urinary bladder correlates with tumor grade and squamous differentiation, Am J Clin Pathol 120(1):93-100.

M. Garber, O. Troyanskaya, K. Schluens, S. Petersen, Z. Thaesler, M. Gengelbach, M. van de Rijn, G. Rosen, C. Perou, R. Whyte, R. Altman, P. Brown, D. Botstein, and L. Petersen (2001) Diversity of gene expression in adenocarcinoma of the lung, PNAS, 98(24): 13784-13798.

K. Hibli, B. Trink, M. Patturajan, W. Westra, O. Caballero, D. Hill, E. Ratovitski, J. Jen and D. Sidransky (2000) AIS is an oncogene amplified in squamous cell carcinoma, PNAS, 97: 5462-5467.

L. Li, C. Wienberg, T. Darden, and L. Pedersen (2001) Gene selection for sample classification based on gene expression data: Study of sensitivity to choice of parameters of the GA/KNN method, Bioinformatics, 17(12): 1131-1142.

L Li, P. Bushel, L. Pedersen, T. Darden, H. Hamadeh, L. Bennett, C. Afshari, R. Paules, D. Umbach, and C. Weinberg (2001) Computational analysis of Leukemia microarray expression data using the GA/KNN method and other existing tools, in Methods of Microarray Data Analysis: Papers from CAMDA 2000, S. Lin and K. Johnson, eds., Boston: Kluwer Academic Publishers.

L. Lin, E. Jeanclos, M. Treuil, K. Braunewell, E. Gundelfinger, R. Anand (2002) The calcium sensor protein visinin-like protein-1 modulates the surface expression and agonist sensitivity of the alpha 4beta 2 nicotinic acetylcholine receptor. J Biol Chem 1 ;277(44): 41872-8.

J. D. Minna (2003) Nicotine exposure and bronchial epithelial cell nicotinic acetylcholine receptor expression in the pathogenesis of lung cancer. J Clin Invest. 111(1): 31-33.

S. Markowitz, J. Wang, L. Myeroff, R. Parson, L. Sun, J. Lutterbaugh, R. Fan, E. Zborowska, K. Kinzler, B. Vogelstein, M. Brattain, and J. Wilson (1995) Inactivation of the type II TGF-beta receptor in colon cancer cells with microsatellite instability, Science, 268: 1336-1338.

D. Slonim (2002) From patterns to pathways: gene expression data analysis comes of age, Nature Genetics, 32(S), 502-508.

M. van de Vijver, J. Petersen, W. Mooi, P. Wisman, J. Lomans, O. Dalesio and R. Nusse (1988) NEU-protein overexpression in breast-cancer: association with comedo-type ductal carcinoma in situ and limited prognostic value in stage II breast cancer, New England Journal of Medicine, 319: 1239-1245.

T. Yang-Feng, A. Schechter, R. Weinberg, U. Francke (1985) Oncogene from rat neuro/glioblastomas (human gene symbol NGL) is located on the proximal long arm of human chromosome 17 and EGFR is confirmed at 7pl3-q11.2 Cytoget. Cell Genetics 40: 784.

W. Zhou, W. Wu, N. Palmer, E. Mower, N. Daniels, L. Cowen, A. Blumer (2004) Microarray data analysis of survival times of patients with lung adenocarcinomas using ADC and K-medians clustering, in Methods of Microarray Data Analysis IV , J. Shoemaker and S. Lin, eds., Boston: Kluwer Academic Publishers (in press).

Chapter 15

MICROARRAY DATA ANALYSIS USING NEURAL NETWORK CLASSIFIERS AND GENE SELECTION METHODS

Gaolin Zheng[1], E. Olusegun George[2], Giri Narasimhan[1,3]
[1]*School of Computer Science, Florida International University, Miami, FL 33199.*
[2]*Mathematical Sciences Department, University of Memphis, Memphis, TN 38152.*
[3]*Corresponding Author.*

Abstract: Different research groups have conducted independent gene expression studies on tissue samples from human lung adenocarcinomas [Bhattacharjee et al. 2001; Beer et al. 2002]. In this paper we (a) investigate methods to integrate data obtained from independent studies, (b) experiment with different gene selection methods to find genes that have significantly differential expression among different tumor stages, (c) study the performance of neural network classifiers with correlated weights, and (d) compare the performance of classifiers based on neural networks and its many variants on gene expression data. Raw cell intensity data were preprocessed for our analyses. Affymetrix array comparison spreadsheets were used to extract the overlapping probe sets for the data integration study. We considered neural network classifiers with random weights selected from a univariate normal distribution and optimized using Bayesian methods. The performance of the neural network was further enhanced using ensemble techniques such as bagging and boosting. The performance of all the resulting classifiers was compared using the Michigan and Harvard data sets from the CAMDA website. Three gene selection methods were used to find significant genes that could discriminate between the various stages of lung cancer. Significant genes, which were mined from the Gene Ontology (GO) database using the GoMiner and AmiGO packages, were found to be involved in apoptosis, angiogenesis, and cell growth and differentiation. Neural networks enhanced with bagging exhibited the best performance among all the classifiers we tested.

Key words: Microarray, lung adenocarcinoma, robust multiarray averaging, gene selection. neural network classifiers, gene ontology

1. INTRODUCTION

Human lung cancer is a major public health problem. More recently, different research groups have conducted independent and systematic microarray-based gene expression studies on a large number of human lung cancer tissue samples [Bhattacharjee et al., 2001; Beer et al., 2002]. The objectives of this paper are (a) to investigate methods to integrate data obtained from independent studies, (b) to experiment with different gene selection methods to find genes that have significantly differential expression among different tumor stages, (c) to study the performance of neural network classifiers with correlated weights when applied to human lung adenocarcinoma gene expression data, and (d) to compare the performance of classifiers based on neural networks and its many variants on the same data.

Data integration is necessary because often, different laboratories, possibly using different microarray technologies and different probe designs, carry out independent investigations. The experiments are expensive and tumor tissues are a precious research resource. It is possible to gain more insight by integrating all the information carefully.

Gene selection methods are important in order to identify critical genes that deserve further biological investigations. They also are useful to reduce the size of the computational problem that is faced when handling enormous microarray data sets.

Classifiers for microarray data for lung cancer tissue samples, if efficacious, can be as a clinical tool (a) to decide whether a new lung tissue sample is cancerous or not, (b) to identify the type of lung cancer, (c) to identify the stage and progress of the disease, and (d) to predict prognosis and survival information about the patient. Classifiers also help to model the data and to identify hidden correlations in them.

Once a list of differentially expressed genes is generated from the microarray data, it is important to understand the relationships among the genes in question. The Gene Ontology (GO) Consortium [Ashburner et al., 2000] maintains databases that help to obtain biological and functional annotations of these genes. GO organizes genes into hierarchical categories based on biological process, molecular function and subcellular localization. Two mining tools AmiGO [www.godatabase.org] and GoMiner [Zeeberg et al. 2003] were used in this study to obtain functional annotations of the significant genes. All the experiments were performed with implementations using the R statistical package [www.cran.r-project.org].

2. DATA ANALYSIS

2.1 Preprocessing

For our analysis, we started with Affymetrix raw cell intensity data. Bioconductor Affy package [www.bioconductor.org] was used to read cell intensity files. All the image files were obtained and the chips with remarkable spatial artifacts were removed from the study.

The popular methods to obtain expression values from Affymetrix cell intensity files are MAS 4.0 AvDiff [www.affymetrix.com], MAS 5.0 Signal [www.affymetrix.com], Li and Wong's Model-Based Expression Index (MBEI) [Li et al. 2001], and robust multiarray averaging (RMA) [Irizarry et al. 2003]. RMA uses only background-corrected perfect match (PM) values, followed by probe level normalization and robust multiarray averaging. RMA was the method chosen for this study because it gives the best summary of bias, variance, and model fit [Irizarry et al. 2003].

2.2 Data integration

We used two data sets described by Bhattacharjee et al. [2001] and Beer et al. [2002]. We refer to the two data sets as the Harvard data sets and the Michigan data sets, respectively. The two studies used different types of Affymetrix chips for their experiments. The Michigan study used the HuGeneFL type chips, while the Harvard study used the HG_U95Av2 type chip. Array Comparison Spreadsheet HuGeneFL to Human Genome U95A [www.affymetrix.com/support] was used to obtain a list of probe sets with 5 or more overlaps for the two Affymetrix chip types. Cell intensity files were read into an AffyBatch object. Invariant set normalization was then performed at the probe level for the AffyBatch object followed by RMA to obtain the expression values. Expression values of the selected probe sets were extracted from both Michigan and Harvard data sets and combined after matching their IDs using the Array Comparison Spreadsheet mentioned above.

2.3 Gene selection

We were interested in identifying genes that could discriminate advanced tumor stages from early tumor stages. Analysis of variance (ANOVA), significance analysis of microarrays (SAM) and a robust gene selection method referred to as GS-Robust, proposed by us, were the three gene selection methods employed in this study.

For the ANOVA model on the data from the individual studies, stage, gender and smoking information were used as fixed factors. For the model on the integrated data, stage, gender and smoking information were used as fixed factors, while the study (i.e., Harvard vs. Michigan) was used as a random factor. Genes were ranked based on their P-values.

Significance analysis of microarrays (SAM), developed by Tusher et al. [2001], was also used to identify significant genes from microarray data. It is more accurate (lower false discovery rates) than conventional methods [Singhal et al., 2003].

GS-Robust was proposed by us as a robust variant of the F-ratio used in ANOVA. Like F-ratio, it too is a measure of the ratio of between groups and within group variations. Larger GS-Robust values indicate higher discrimination power. For the ith gene, the GS-Robust statistic is defined by

$$GSRobust_i = \frac{MAD[median(\underline{g_{i1}}),...,median(\underline{g_{ik}})]}{\sum_{j=1}^{k} MAD(\underline{g_{ij}})} \quad (1)$$

where $\underline{g_{ij}}$ is the vector of gene expression values for the ith gene in the jth class, and k is the total number of classes. Unlike F-ratio, GS-Robust uses median absolute deviation, and substitutes mean with median measures. GS-Robust is, therefore, less sensitive to outliers. A disadvantage of the GS-Robust statistic is that it does not have a standard null distribution. As such statistical significance (p-values) may be evaluated by using a bootstrap or permutation resampling procedure. Another disadvantage of GS-Robust (and also SAM) is that there is no obvious approach to extend it to models with multiple factors. However the degrees of freedom for the statistic are the same for all the genes, we can use this measure to rank the discriminative power of the genes. In this paper, a comparative study was performed on the three gene selection methods mentioned above.

Principal component analysis (PCA), a data reduction method, was also used in this study to select the desirable input features for classification. PCA was performed on the correlation matrix. As is customary, in measurements that have different scales, we used the correlation matrix because of the intrinsic heteroscedastic nature of gene expression. Moreover, although principal components are not scale invariant, the principal components generated from correlation matrices are more tractable and allow for more meaningful comparisons of genes. The principal components contributing to at least 75% of the variation were used for classification.

2.4 Neural network classifiers

A neural network implements a non-linear function $y(x, w)$, where y is the output function for input x and network parameters (or weights) w. Given a training set, i.e., set of pairs of the form $\langle x_i, y_i \rangle, i = 1,\ldots, N$, the neural network can be trained to model the given data as closely as possible, and thereby determine the weight vector w that best describes the given data. The training procedure involves minimizing an appropriate error function. Once the optimal weight vector is determined, the neural network acts as a classification or regression tool, depending on whether the output is from a discrete or continuous set of values. For the sake of comparison, support vector machines (SVM), K nearest neighbor (KNN), and random forest classifiers were also implemented and tested.

Neural networks have been used to model gene expression data, where the output function may represent a medical condition or some clinical or biological event such as the recurrence of a disease or prognosis of certain cancers [Khan et al., 2001; Ando et al., 2002; Mateos et al., 2002; Grey et al., 2003]. However in these papers, the network parameters w are assumed fixed deterministic constants.

In such models where the weights are not random, the correlations that exist between outputs are artificially induced through the iterative process of the neural network itself. However, these correlations need to be explicitly incorporated into the model. One way to do this is through weight vector (network parameters). Using random weight components induces correlation among genes, since the posterior weights become correlated and account for the fact that genes act in concert with a collection of other genes forming gene networks. In this paper we assume a simple correlation model, i.e., that components of the weight vector are random under a univariate model.

2.4.1 Bayesian regularization of network weights

In regular neural networks, after initializing the network parameters by choosing randomly from a univariate model, the training set is used to optimize the network parameters. The method can be further improved by determining the parameters of the univariate model using standard Bayesian techniques. This is achieved by choosing the optimal weights as the modes of the posterior probability density functions $P(w|\langle x_i, y_i\rangle)$, i.e., by maximizing $P(w|\langle x_i, y_i\rangle)$. Here $P(w|\langle x_i, y_i\rangle)$ is the posterior probability of network weights given the input data. In this paper, we report on experiments comparing the performance of regular neural networks to that of its Bayesian counterpart using lung cancer gene expression data.

2.4.2 Ensemble techniques

More recently, it has been shown that using ensemble techniques such as bagging and/or boosting can enhance the performance of classifiers. Both these techniques are termed as "ensemble" techniques because they correspond to designing a "committee" of classifiers such that their collective performance surpasses their individual performance.

Bagging: Bagging is an acronym for "bootstrap aggregating" [Breiman 1996]. The idea is to design k data sets (denoted by $D_1, D_2, \ldots, D_k$) by a process of repeated bootstrap sampling from the original data set, and to design k independent classifiers using them as the training sets. For any given test data, all the k classifiers vote to give a resulting classification. Breiman has noted that neural network classifiers tend to be unstable [Breiman 1996], and that bagging tends to improve unstable classification methods more than stable ones. In this paper, we report on experiments comparing the performance of regular neural networks and their Bayesian counterparts with and without bagging.

Boosting: Boosting was designed to boost the performance of weak classifiers [Schapire 1990]. As in bagging, k classifiers are successively designed. Unlike with bagging, the training samples are weighted with all samples having equal weights initially. In successive classifiers, weights are iteratively modified so that higher weights are assigned to samples misclassified in previous classifiers and the expected error over different input distributions is minimized. After the classifiers are designed, they are assigned weights based on their performance on the training data. A weighted voting scheme is then used to determine the resulting classification for a given test sample. In this paper, we report on experiments comparing the performance of regular neural networks and their Bayesian counterparts with and without the enhancement of boosting.

2.4.3 K-fold cross-validation

In order to compare the performance of the various classifiers mentioned above, we used the standard statistical method of K-fold cross-validation. According to this method, the data was divided into K groups and K separate tests were run. When testing samples from each of the groups, the classifier was trained with the K-1 remaining groups. The error rate was reported after averaging over all the groups.

2.4.4 Practical issues

When designing classifiers for data sets with two or more categories, the training data set may not be balanced in the sense that the number of samples in each category may not be the same. This may cause a bias in the classifiers that are designed. To address this problem, one could create bootstrap copies of samples from the underrepresented classes until a balance is achieved [Japkowicz 2000], or one could randomly remove samples from the overrepresented classes. The first approach suffers from oversampling. The second approach tends to lose potentially significant information. Choosing the lesser of the two evils, we adopted the first approach to adjust the classifiers. It was not used in our gene selection study.

3. RESULTS AND DISCUSSIONS

3.1 Preprocessing

Five of the chips from the Michigan data set, namely L01, L54, L88, L89, and L90, had remarkable spatial artifacts (Figure 1), and were removed from the study. Data on 81 patients were used in the study, of which 64 patients had stage 1 adenocarcinoma, and 17 had stage 3 adenocarcinoma. Of the 81 individuals, 48 were women and 33 were men. Only eight of the 81 were non-smokers while the rest were smokers. Gene expression values for the 7129 probe sets were generated using RMA.

In the Harvard data set, four chips, namely CL2001032701AA, CL2001032709AA, CL2001032634AA, and CL2001032623AA had remarkable spatial artifacts (Figure 1) and were removed from the study. Expression values from the replicates were averaged. After this step, 76 stage 1 adenocarcinoma tumor samples, 24 stage 2 adenocarcinoma tumor samples, and 10 stage 3 adenocarcinoma tumor samples were used for our analyses. Sixty-five of the patients were women, and 45 samples were men. Only 12 of the 110 were nonsmokers. Expression data for the 12625 probe sets were generated using RMA.

To produce an integrated data set from the two data sets, we chose 3742 probe sets that had five or more overlaps in the two data sets. The overlap information was obtained using the Array Comparison Spreadsheet available on the Affymetrix website [www.affymetrix.com/support]. Corresponding subsets of data (corresponding to the 3742 chosen probe sets) from the Michigan and Harvard studies were also generated for our experiments on individual data sets. The perfect match and mismatch intensities from the subsets were normalized using invariant set separately. Robust multiarray averaging method was applied to the AffyBatch object resulting from invariant set normalization to generate the expression values for the 3742 probe sets.

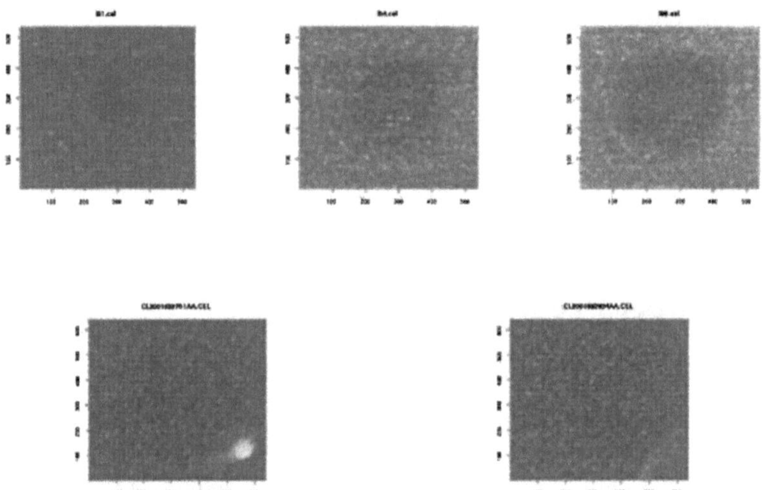

Figure 1. Images of the chips from the Michigan (top row) and Harvard (bottom row) data sets with remarkable spatial artifacts.

3.2 Identifying genes discriminating the tumor stages

Three lists of top 500 genes were generated using multifactor ANOVA, GS-Robust and SAM. Figure 2 shows the intersections of the three groups.

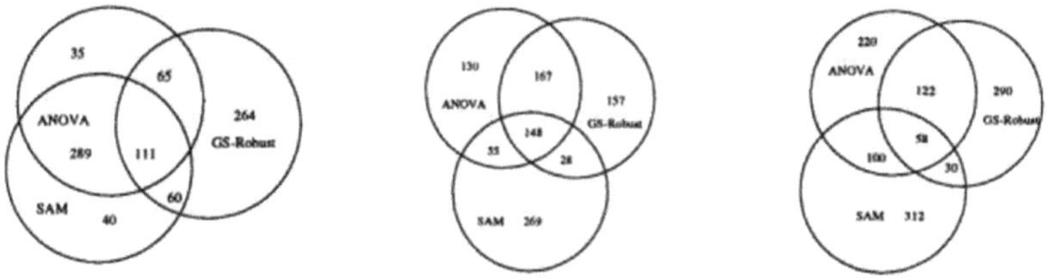

(a) (b) (c)

Figure 2. The intersection of the top 500 genes obtained using the three gene selection methods on the (a) Harvard, (b) Michigan, and (c) integrated data sets.

GS-Robust (for the Michigan data set) and the SAM method (for the Harvard data set) selected a list of significant genes that were considerably different from the ones picked by the other methods. For the integrated data set, the overlap in the top 500 lists generated by the three methods was greatly reduced.

3.2.1 Querying significant genes against GO Database

The Gene Ontology (GO) database was queried using GoMiner [Zeeberg et al., 2003]. Significant genes selected using ANOVA were fed into GoMiner and p-values were computed for each GO term based on Fisher's exact tests [Zeeberg et al., 2003] as follows: Let p_1 be the probability that a gene will be flagged under the GO term and p_2 be the probability that it will not. The null hypothesis H_0: $p_1 = p_2$, will be true if genes are flagged under the GO term purely by chance, and there is no significant difference in the two categories. We use the Fisher's exact test to test this hypothesis. This is a conditional test given the sufficient statistics $(n_f/n,(N_f-n_f)/(N-n))$ where n_f is the number of flagged genes under the GO term, n is total number of genes under the GO term, N_f is number of flagged genes on the microarray, and N is the total number of genes on the microarray.

Identifying significant genes: With the help of GoMiner and the Unigene Ids, some of the significant genes (for each of the three sets) and the biological process they are involved in are given below in Table 1.

The analysis of the Michigan data set resulted in five molecular function (MF) GO terms (and their relationships) with p-value less than 0.01 (see Figure 3). A similar analysis of the Harvard data set resulted in 12 MF GO terms (and their relationships), as shown in Figure 4.

Finally, an analysis of the integrated data set gave 6 MF GO terms (and their relationships), as shown below in Figure 5.

Table 1. Significant genes identified from the three data sets.

Study	***Biological Process***	***Induced***	***Repressed***
Michigan	Apoptosis	BIRC2	BBC3, MUC2, PLG
	Angiogenesis	FGF2, POFUT1, VEGF	EPAS
	Cell growth		TGFB
	Cell Cycle	CDC27, CDC7, CDK7, CKS2	
Harvard	Apoptosis	PRKAA1, GSK3B	CASP3, PLG
	Angiogenesis	VEGF	
	DNA Replication	DNTT, SSBP1	
Integrated	Apoptosis		CASP3
	Angiogenesis	VEGFC	
	Cell differentiation	MYF5, PAX6	

Figure 3. Relationships among significant GO terms identified from the Michigan data set. Note that the significant terms with more overexpressed genes (dark circles), more underexpressed genes (gray circles), and with insignificant changes (white circles) are marked appropriately.

Figure 4. The relationships among the significant MF GO terms identified from the Harvard data set. Note that the significant terms with more overexpressed genes (dark circles), more underexpressed genes (gray circles), equal number of overexpressed and underexpressed genes (dotted circles), and with insignificant changes (white circles) are marked appropriately.

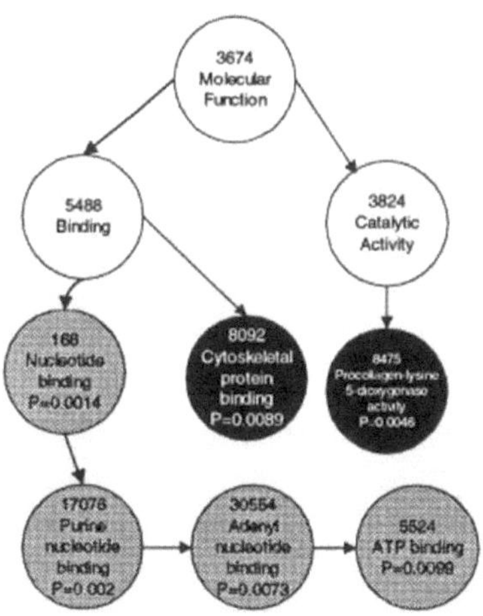

Figure 5. The relationships among the significant MF GO terms identified from the integrated data set. Note that the significant terms with more overexpressed genes (dark circles), more underexpressed genes (gray circles), and with insignificant changes (white circles) are marked appropriately.

3.3 Classification results

Tables 2, 3, and 4 show the results from our experiments with neural network classifiers using stage information from the Michigan, Harvard, and integrated data sets. Additional classifiers such as SVM, KNN, and random forests were also used for comparisons purposes. In all three sets of experiments, genes were selected using three different ranking schemes and PCA, and the results shown are the mean ± SD of 5-fold cross-validation error from 10 independent runs.

Table 5 shows the results of our cross-validation experiments. When trained with the Michigan data set and tested with the Harvard data set, an accuracy of up to 88% was achieved using bagged neural network classifiers with genes selected using ANOVA. When the roles of the data sets were reversed, an accuracy of only 80% was achieved with most of the gene selection and bagged neural network classifiers. Note that the Michigan data set did not have any data from patients with stage 2 tumors. Only stages 1 and 3 (T1 and T3) were available. Therefore, when we trained with the Michigan data set, all stage 2 data from the Harvard set were left out of the testing. However, when we trained with the Harvard data set, data from all the stages was used (T1, T2 and T3).

Table 2. Experiments on NN classifiers on stage information from the Michigan data set.

	Gene Selection Methods			
	ANOVA	SAM	GS-Robust	GS-PCA
nnet	18.5±3.2%	30.8±6.2%	20.0±2.7%	18.5±2.0%
nnet.bag	16.9±2.7%	23.5±2.6%	18.0±2.1%	14.7±3.1%
nnet.boost	19.7±2.2%	29.2±8.0%	18.8+2.4%	21.2±4.4%
bayesian	15.1± 2.8%	42.3±6.7%	18.3±3.1%	17.2±3.5%
bayes.bag	14.1±2.8%	30.9±2.0%	18.4±2.3%	14.0±2.8%
bayes.boost	17.3±2.4%	38.7±4.1%	19.2±2.4%	17.1±3.0%
SVM	21.4±0.6%	20.8±1.4%	20.5±1.0%	21.5±0.4%
KNN	25.3±0.0%	26.7±0.0%	18.7±0.0%	25.3±0.0%
RandomForest	24.7±0.7%	19.6±0.9%	18.5±1.3%	20.4±1.7%

Table 3. Experiments on NN classifiers on stage information from the Harvard data set.

	Gene Selection Methods			
	ANOVA	SAM	GS-Robust	GS-PCA
nnet	14.6±2.4%	14.0±2.9%	17.7±5.6%	15.1±3.1%
nnet.bag	12.2±1.6%	12.4±1.0%	13.8±2.4%	12.5±3.3%
nnet.boost	14.2±3.0%	15.4±3.1%	18.3±3.3%	19.8±5.7%
bayesian	17.1±2.7%	14.9±2.5%	20.8±3.3%	21.0±4.5%
bayes.bag	12.9±2.2%	13.6±1.8%	17.1±1.8%	18.2±2.1%
bayes.boost	17.1±3.0%	16.1±2.5%	21.3±2.6%	23.3±2.3%
SVM	19.0±0.0%	19.0±0.3%	18.9±0.0%	19.6±0.4%
KNN	21.8±1.3%	22.7±1.0%	13.4±1.3%	29.2±1.5%
RandomForest	17.9±0.7%	17.7±0.1%	18.7±0.1%	20.3±1.1%

Table 4. Experiments on NN classifiers on stage information from the integrated data set.

	Gene Selection Methods			
	ANOVA	SAM	GS-Robust	GS-PCA
nnet	13.1±2.0%	17.4+1.9%	12.4±1.7%	13.6±2.5%
nnet.bag	11.3±1.1%	13.3±1.6%	9.3±1.4%	12.1±0.8%
nnet.boost	13.3±2.9%	18.8±4.8%	11.5±2.1%	15.4±3.9%
bayesian	16.7±2.9%	18.8±4.7%	10.9±2.6%	15.1±2.2%
bayes.bag	14.2±2.4%	24.7±2.4%	10.6±2.2%	14.6±2.6%
bayes.boost	16.7±4.8%	19.3±5.1%	13.1±3.2%	17.1±5.5%
SVM	14.8±0.7%	15.2±0.4%	14.2±0.2%	14.5±0.7%
KNN	18.5±0.5%	15.1±0.9%	10.9±0.6%	18.1±0.9%
RandomForest	14.4±0.7%	14.7±0.6%	14.8±0.9%	14.4±1.1%

Table 5. Cross-validation experiments.

Training Set	Testing Set	Classifier Method	Gene Selection Methods		
			ANOVA	SAM	GS-Robust
Michigan (T1 and T3)	Harvard (T1, T3)	nnet	39.5±4.9%	28.7±4.1%	25.9±1.4%
		nnet.bag	11.6±3.3%	20.0±5.6%	13.9±5.7%
		nnet.boost	17.4±4.7%	22.4±4.7%	21.7±8.9%
		Bayesian	18.8±3.0%	25.0±5.3%	26.6±6.7%
		bayes.bag	12.8±0.1%	21.0±0.5%	20.1±0.6%
		bayes.boost	25.5±0.4%	29.8±1.7%	28.9±1.5%
		SVM	14.7±0.3%	15.2±0.4%	14.2±0.2%
		KNN	18.5±0.5%	15.1±0.9%	18.1±0.9%
		RandomForest	14.4±0.7%	14.7±0.6%	14.4±1.1%
Harvard (T1, T2, T3)	Michigan (T1 and T3)	nnet	27.4+17.8%	21.0±0.5%	21.0±0.6%
		nnet.bag	22.3±4.3%	20.9±0.3%	21.1+0.4%
		nnet.boost	42.3±25.5%	21.0±0.5%	26.7±18.0%
		Bayesian	33.7±17.9%	22.2±3.9%	21.0±0.1%
		bayes.bag	32.3±23.0%	20.9±0.7%	21.2±0.5%
		bayes.boost	33.7±14.3%	21.1±0.4%	21.1±0.7%
		SVM	29.0±0.2%	24.4±0.3%	20.3±0.3%
		KNN	29.9±3.1%	23.6±1.5%	21.9±3.2%
		RandomForest	30.7±5.0%	22.3±2.7%	20.4±1.7%

4. CONCLUSIONS

Bagging consistently and significantly improved the performance of feed-forward neural network classifiers in all our experiments. Since bagging incurs only a small amount of computational overhead, it is feasible to apply this ensemble technique to enhance most classifiers. Boosting, on the other hand, showed erratic behavior. Bayesian neural networks did not show any appreciable improvement over the regular neural networks.

The performance of all the gene selection methods was comparable, with two exceptions. It was not clear why SAM performed poorly only on the Michigan data set. GS-Robust performed particularly well on the integrated data set. We conjecture that GS-Robust was better able to cope with the extra noise that must have been introduced during the data integration process. With gene expression data preprocessed using a robust method such as RMA, the performance of ANOVA and GS-Robust were comparable. Without RMA, GS-Robust outperformed ANOVA (data not shown).

Genes significant for carcinoma stage differentiation were identified from the Michigan, Harvard, and the integrated data sets based on our results from analysis of variance at a significance level of 0.05. Among the significant genes identified from the Michigan data set were three apoptosis activators that were repressed significantly, while one apoptosis inhibitor was induced significantly (Table 1). Interestingly, several cell cycle genes (CDC27, CDC7, CDK7, and CKS2) were induced. In contrast, the cell growth gene TGFB, which is related to lung development, was repressed. Three angiogenesis genes were induced significantly, while only one angiogenesis gene was repressed.

In the advanced stage tumors in the Harvard data set, apoptosis activators, PLG and CASP3, were repressed, while apoptosis inhibitors, PRKAA1 and GSK3B, were induced. Genes involved in DNA replication (DNTT and SSBP1), and the angiogenesis-related gene, VEGF, were induced significantly in the advanced stage tumors of the Harvard data set.

Cell differentiation genes, MYF5 and PAX6, were induced significantly in advanced stage tumors of the integrated data set, as did the angiogenesis-related gene, VEGFC. In contrast, CASP3 (which was also identified from the Harvard data set) was repressed. In summary, genes PLG (apoptosis), CASP3 (apoptosis), and VEGF (angiogenesis) were identified as significant from two independent data sets.

Acknowledgements

Research of E.O.G. & G.N. was supported by NIH Grant P01 DA15027-01.

REFERENCES

Ando, T., M. Suguro, T. Hanai, T. Kobayashi, H. Honda and M. Seto (2002). "Fuzzy neural network applied to gene expression profiling for predicting the prognosis of diffuse large B-cell lymphoma." Japanese Journal of Cancer Research 93(11): 1207-12.

Ashburner, M., C. A. Ball, J. A. Blake, D. Botstein, H. Butler, J. M. Cherry, A. P. Davis, K. Dolinski, S. S. Dwight, J. T. Eppig, M. A. Harris, D. P. Hill, L. Issel-Tarver, A. Kasarskis, S. Lewis, J. C. Matese, J. E. Richardson, M. Ringwald, G. M. Rubin and G. Sherlock (2000). "Gene Ontology: tool for the unification of biology." Nature Genetics 25: 25 - 29.

Beer, D. G., S. L. R. Kardia, C.-C. Huang, T. J. Giordano, A. M. Levin, D. E. Misek, L. Lin, G. Chen, T. G. Gharib, D. G. Thomas, M. L. Lizyness, R. Kuick, S. H. Hayasaka, J. M. G. Taylor, M. D. Iannettoni, M. B. Orringer and S. Hanash (2002). "Gene-expression profiles predict survival of patients with lung adenocarcinoma." Nature Medicine 8(8): 816-24.

Bhattacharjee, A., W. G. Richards, J. Staunton, C. Li, S. Monti, P. Vasa, C. Ladd, J. Beheshti, R. Bueno, M. Gillette, M. Loda, G. Weber, E. J. Mark, E. S. Lander, W. Wong, B. E. Johnson, T. R. Golub, D. J. Sugarbaker and M. Meyerson (2001). "Expression profiling reveals distinct adenocarcinoma subclasses." PNAS 98(24): 13790–13795.

Breiman, L. (1996). "Bagging predictors." Machine Learning J. 24(2): 123-40.

Grey, S., S. Dlay, B. Leone, F. Cajone and G. Sherbet (2003). "Prediction of nodal spread of breast cancer by using artificial neural network-based analyses of S100A4, nm23 and steroid receptor expression." Clin Exp Metastasis 20(6): 507-14.

Irizarry, R., B. Hobbs, F. Collin, Y. Beazer-Barclay, K. Antonellis, U. Scherf and T. Speed (2003). "Exploration, normalization, and summaries of high density oligonucleotide array probe level data." Biostatistics 4(2): 249-264.

Japkowicz, N. (2000). Class imbalance problem: significance and strategies. International Conference on Artificial Intelligence (IC-AI'2000): Special Track on Inductive Learning, Las Vegas.

Khan, J., J. S. Wei, M. Ringner, L. H. Saal, M. Ladanyi, F. Westermann, F. Berthold, M. Schwab, C. R. Antonescu, C. Peterson and P. S. Meltzer (2001). "Classification and diagnostic prediction of cancers using gene expression profiling and artificial neural networks." Nat Med 7(6): 673-9.

Li, C. and W. H. Wong (2001). "Model-based analysis of oligonucleotide arrays: Expression index computation and outlier detection." PNAS 98(1): 31-36.

Mateos, A., J. Herrero, J. Tamames and J. Dopazo (2002). Supervised Neural Networks for Clustering Conditions in DNA Array Data after Reducing Noise by Clustering Gene Expression Profiles. Methods of Microarray Data Analysis II. S. M. Lin and K. F. Johnson. Boston, Kluwer Academic Publishers.

Schapire, R. E. (1990). "The strength of weak learnability." Machine Learning J. 5(2): 197-227.

Singhal, S., C. G. Kyvernitis, S. W. Johnson, L. R. Kaiser, M. N. Liebman and S. M. Albelda (2003). "MicroArray Data Simulator For Improved Selection of Differentially Expressed Genes." Cancer Biology & Therapy 2(4): 383-391.

Tusher, V. G., R. Tibshirani and G. Chu (2001). "Significance analysis of microarrays applied to the ionizing radiation response." PNAS 98(9): 5116-5121.

Zeeberg, B. R., W. Feng, G. Wang, M. D. Wang, A. T. Fojo, M. Sunshine, S. Narasimhan, D. W. Kane, W. C. Reinhold, S. Lababidi, K. J. Bussey, J. Riss, J. C. Barrett and J. N. Weinstein (2003). "GoMiner: A Resource for Biological Interpretation of Genomic and Proteomic Data." Genome Biology 4(4): R28.

Chapter 16

A COMBINATORIAL APPROACH TO THE ANALYSIS OF DIFFERENTIAL GENE EXPRESSION DATA

*The Use of Graph Algorithms for Disease Prediction and Screening**

Michael A. Langston[1], Lan Lin[1], Xinxia Peng[2], Nicole E. Baldwin[1], Christopher T. Symons[1], Bing Zhang[3] and Jay R. Snoddy[3]

[1]Department of Computer Science, University of Tennessee, Knoxville, TN 37996-3450; [2]Graduate School of Genome Science and Technology, University of Tennessee, Knoxville, TN 37996-0845; [3] Life Sciences Division. Oak Ridge National Laboratory, P.O. Box 2008, Oak Ridge, TN 37831-6124.

Abstract: Combinatorial methods are studied in an effort to gauge their potential utility in the analysis of differential gene expression data. Patient and gene relationships are modeled using edge-weighted graphs. Two algorithms with different, but complementary approaches are devised and implemented. One is based on finding optimal cliques within general graphs, the other on isolating near-optimal dominating sets within bipartite graphs. A main goal is to develop methodologies for training algorithms on patient populations with known disease profiles, so that they can be employed to classify and predict the likelihood of disease in patient populations whose profiles are not known. These novel strategies are in marked contrast with Bayesian and other well-known techniques. Encouraging results are reported.

Key words: Combinatorial methods; discrete mathematics; disease prediction and screening; graph algorithms; graph theory; microarray analysis

* This research is supported in part by the National Science Foundation under grants EIA-9972889, CCR-0075792 and CCR-0311500, by the Office of Naval Research under grant N00014-01-1-0608, by the National Institutes of Health under grant U01-AA013512-02, by the Department of Energy under contract DE-AC05-00OR22725, and by the Tennessee Center for Information Technology Research under award E01-0178-261.

1. INTRODUCTION

A fundamental problem in cancer treatment is early and reliable detection. Identification of a set of genes whose expression levels serve as an accurate discriminator among normal and cancerous tissue samples would not only represent significant progress towards developing more reliable cancer diagnosis protocols, but might also identify novel therapeutic targets. With this motivation in mind, we investigated the hypothesis that only a modest number of genes may suffice for this task. We sought to develop algorithms and software for this purpose, and introduced a graph theoretical method of differential gene expression analysis. The goals of this method were to identify a set of genes useful in discriminating among tissue samples, and to use these genes in disease prediction and screening.

One of the important features of our algorithms was the computation of discrimination scores for each gene represented in a microarray. These scores estimated a gene's relative ability to distinguish among sample tissue classes. We then selected the highest-scoring genes, and used them to calculate a pairwise similarity metric between patients' tissue sample expression profiles. Genes that failed to discriminate among a defined percentage of the samples were eliminated using a dominating set algorithm as a high pass filter. With this information, we constructed a complete weighted graph, in which the vertices represent the tissue samples and the edges are weighted by the similarity metric between sample vertices. A user-defined threshold was then used to transform the complete weighted graph into an incomplete unweighted graph where the weights were ignored. The combination of these tools produced some very encouraging predictive results.

In the sequel, we describe the datasets we chose to study, the algorithms we devised, and the results we obtained. We also draw some conclusions from this effort.

2. DATA EMPLOYED

We used the Harvard [Bhattacharjee et al., 2001.], Michigan [Beer et al., 2002], and Stanford [Garber et al., 2001] datasets in this study. We did not include the Ontario dataset due to a lack of overlap in annotated genes with the other datasets. Since the log-expression image plots for Samples L54, L88, L89 and L90 in the Michigan dataset showed large, round dark spots at the center of the arrays [Hu et al., 2003] indicative of poor data quality, they were removed from the dataset. This left us with 92 samples from the Michigan dataset. Because the Harvard and Michigan datasets were generated by different institutes using different Affymetrix array types (HG_U95A and HUGeneFL, respectively), the distributions of the two datasets may not be comparable. Thus, we chose to normalize the two datasets separately. The log-scale quantifications of the gene expression levels for each probe set were obtained by robust multi-array average (RMA) [Irizarry et al., 2003] using Bioconductor.

Since we intended to train and test our algorithms on different datasets, we needed a mapping schema among the different datasets. However, the three datasets came from different array platforms using different gene identifiers; hence, direct mapping is not possible. We chose to use LocusLink IDs (LL_IDs) for gene mapping, because the NCBI LocusLink Database is both relatively reliable and stable. For the Harvard and Michigan datasets, we mapped each probe set ID to its corresponding LL_ID using array annotation files from Affymetrix. For the Stanford dataset, we mapped each UNIGENE ID to its corresponding LL_ID using our local database, GeneKeyDB. To construct a gene expression summary for each LL_ID, we averaged the values within each sample across the original gene identifiers that map to a common LL_ID. The final datasets used in this study include: the Harvard dataset, which has expression profiles for 8509 unique genes among 254 samples; the Michigan dataset, which has expression profiles for 4985 unique genes among 92 samples; and the Stanford dataset, which has expression profiles for 8829 unique genes among 73 samples.

3. A CLIQUE-BASED STRATEGY

3.1 The Clique Problem

Clique is a well-known *NP*-complete problem (informally, this means that the best solution procedures possible seem to require time exponential in the size of the input), and is typically formulated as in Garey and Johnson [1979]:

Input: A graph $G=(V,E)$ and a positive integer $k \leq |V|$.

Question: Is there a subset $V' \subseteq V$ for which $|V'| \geq k$ and such that every pair of vertices in V' is joined by an edge in E.

Thus, a clique is a subgraph each of whose nodes is pairwise related. Clique is rapidly becoming recognized for its relevance in bioinformatics. It can be roughly viewed as a clustering algorithm based on graph theory. In our own work, for example, we used clique in the following ways. In Abu-Khzam et al. [2003], we devised and applied fast parallel algorithms for clique to extremely large microarray datasets in an effort to help identify putatively co-regulated genes in murine neural regulatory networks. In another application [Baldwin et al., 2004], we employed high performance implementations of clique in the study of *cis*-regulatory elements to discover putative motifs.

3.2 Scoring Method

Our goal in training was to develop graph-theoretic tools to help distinguish among sample groups (such as normal and adenocarcinoma). Ideally, we hoped to be able to construct an unweighted graph in which edges connected mainly members of the same group. At that point, clique analysis was an attractive approach for testing our methods against additional data.

In order to pinpoint a modest number of genes out of thousands from the original dataset, our first step in training was to determine which genes appear to discriminate best among sample types. To accomplish this, a discrimination score was calculated for each gene. Only the best genes (those with the highest scores) were retained for subsequent steps. Since the distributions of the expression values of these genes would be expected to be bimodal with respect to two distinct sample classes, the differences between class medians gave us a general measure of the difference of expression between two classes. Subtracting the sum of the standard deviations of a gene within each group allowed us to eliminate, or at least diminish, the importance of any gene whose expression levels vary excessively.

The data was obtained as in Section 2 as an n x m matrix, A, of expression values. Rows represent test samples, and columns denote genes. When training on the Michigan dataset in order to learn to distinguish between normal (group 1) and adenocarcinoma (group 2) samples and using a lower limit of zero, our method delivered a collection of 105 genes for further evaluation.

An assignment of inter-sample weights helped demonstrate the degree to which these genes and their respective scores delineated normal samples from adenocarcinoma. Here, the weight between samples i and j represented the degree of similarity in their respective expression profiles and could be viewed as equivalent to the distance function for clustering. We computed this weight as a sum over all genes selected in the previous step, because it was these genes that seemed to have the greatest potential to serve as good discriminators. Accordingly, we set *weight(i,j)* to:

$$(1) \sum \text{score}(\text{gene}_k) \bullet (1 - |\text{expression\_value}_{ik} - \text{expression\_value}_{jk}|)$$

As is shown in Figure 1, higher-weighted sample pairs tended to be homogeneous. That is, either both tissue samples were normal or both were adenocarcinoma. Conversely, lower-weighted pairs tended to be heterogeneous, where one sample was normal and the other was adenocarcinoma. While this seemed to confirm our gene scoring and selection procedure, other scoring approaches appeared to be viable as well. Therefore, we investigated several other alternatives before settling on this approach.

Two of these alternative approaches are worthy of note in the computation of gene discrimination scores. One is the elimination of outliers before computing the scores, which was motivated by the fact that outliers might affect both the median and the standard deviation. The other involves changing our original scoring function to a variant of the t-test function, a standard statistical measurement of population similarity. This test is realized using division rather than subtraction within our scoring function. Neither of these appeared to improve upon our original results. We also experimented with Pearson's correlation coefficients and Spearman's rank correlation coefficients, two popular methods of weighting. Neither of these methods was helpful. In fact, neither even revealed the bimodal distribution we observed using our weight function.

In addition to confirming the validity of our approach, Figure 1 also suggests an initial threshold weight below which we delete edges in a later step. Call this threshold *T*. For example, based on the figure, we chose as a somewhat informed but still rather arbitrary starting value *T*=7.6. We used our restricted set of genes to build an edge-weighted graph. In this graph, samples were represented by vertices and the weight of an edge between a sample pair was set using the simple summation formula already described.

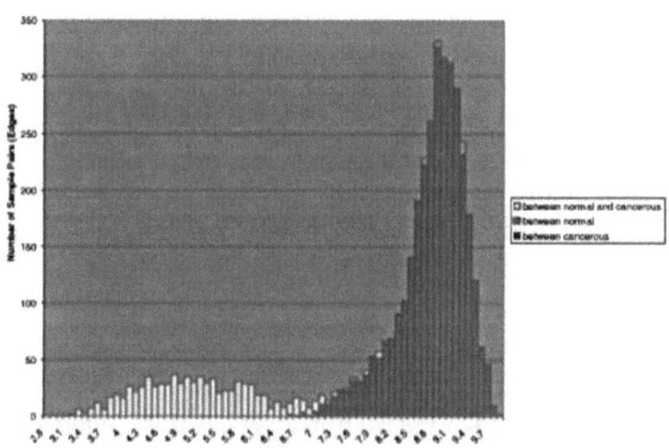

Figure 1. Weights between sample pairs using 105 genes from the Michigan dataset.

Any edge whose weight was less than T was removed. The resulting unweighted graph was then searched for all maximal cliques. Our aim was to train our codes so that we could find appropriately sized cliques to cover all groups.

Because we know which samples are normal and which are adenocarcinoma in the Michigan dataset, we were able to iterate our method until we had a reasonable set of covering cliques. The optimal threshold seemed to be centered at around T=8.1. We were not completely satisfied, however, with the lingering presence of overlapping cliques. Additional experimentation with gene cutoff scores seemed to indicate that the presence of genes with low scores is problematic. But neither raising the cutoff score nor additional modification of the threshold was of much use.

4. REFINEMENT VIA DOMINATING SET

What seemed to be missing in our estimates of gene discrimination was a way to determine which genes impact the greatest number of samples and to eliminate the rest. For this, we turned to another graph metric, dominating set.

4.1 The Dominating Set Problem

Dominating Set, another well-known *NP*-complete problem, can be stated as follows.

Input: A graph $G=(V,E)$ and a positive integer $k \leq |V|$.
Question: Is there a subset $V' \subseteq V$ for which $|V'| \leq k$ and every vertex $v \in V - V'$ is joined to a vertex in V' by an edge in E.

Using the W hierarchy from the theory of fixed-parameter tractablility (FPT), dominating set may be even more difficult than clique. This is because clique, which is W[l]-complete, can be solved using graph complementation and vertex cover. Practical, efficient kernelization techniques are known for vertex cover [Abu-Khzam et al., 2004]. The same, however, may not hold for dominating set. The dominating set version we address here is nonplanar red/blue dominating set, which is W[2]-complete. Although its complement problem is FPT, there are currently no practical kernelization techniques known for it. Thus, we only approximated solutions to dominating set. For technical definitions and discussion of the W hierarchy, see [Downey and Fellows, 1999].

4.2 Scoring Method

We first assumed a normal distribution of the expression values of each gene, and estimated for it the mean and standard deviation. We did this separately for each of the sample groups. Then, based on the estimated normal distribution, we calculated the p-values for the original individual expression values. It is perhaps easiest to formulate our approach by constructing a bipartite graph. In this graph, one set of vertices represents the genes, and the opposing set represents the samples. We placed an edge between a gene and a sample if and only if the p-value of the expression value corresponding to that gene-sample combination was greater than 0.05. Following statistical convention, we considered a p-value below this cutoff to indicate an outlier.

In this setting, we wanted to identify the genes that dominate (or nearly dominate) all the samples. Therefore, we winnowed out from consideration any gene vertex not adjacent to at least 90% of the sample vertices. For example, in the Michigan dataset, a gene was eliminated if it was connected to fewer than 74 of the adenocarcinoma samples or fewer than nine of the normal samples. The choice of 90% was arbitrary; it was selected only after extensive testing.

Next, in an effort to remove any remaining genes with a low possibility of discriminating between the two groups, we calculated the p-values for tests of equal means using both the Wilcoxon and t-test methods. We used both since the t-test assumes a normal distribution, while the Wilcoxon test does not. Only genes for which both p-values are less than 0.05 were retained.

For those genes that remain, we generated scores based on the previously calculated p-values from the Wilcoxon tests. We then filtered out genes using an adjusted p-value cutoff by means of the Bonferroni method. Specifically, we chose a significance level of $\alpha = 0.01$ and only kept genes with a p-value less than α/N, where N is the total number of genes we began with at this step. Since a smaller p-value indicates a greater probability that the groups' expression values are different for a given gene, we used -log10(p-value) for the gene score.

Finally, and most importantly, we computed the intersection of the genes identified by the clique-based approach described in the last section with the genes chosen by the dominating set method as described in this section. We were left with a set of genes that passed both the clique and the dominating set tests. We found that this refinement of our gene lists gave us improved results in the testing phase of our experiments.

5. RESULTS

Having completed the training phase, we proceeded to testing on a new dataset under the assumption that we did not know sample classification in advance. We evaluated our approach with the following three experiments. First, we trained on the Michigan dataset as explained in section 3 in order to learn to distinguish between normal and adenocarcinoma samples. We proceeded to test our ability to classify samples on the Harvard dataset. Second, we reversed this process, applying our training algorithms to the Harvard dataset to distinguish between cancerous and normal samples. We tested our method on the Michigan dataset. Third, we trained on the Harvard dataset to learn to separate adenocarcinoma from squamous samples, and tested on the Stanford dataset.

5.1 Experiment One

Clique-based training on the Michigan dataset identified 105 genes that distinguished between adenocarcinoma and normal samples. Our dominating- set-based refinement reduced this to 84 genes, 78 of which were available in the Harvard data. Functional classification of the selected 84 genes was performed using the web-based tool Gene Ontology Tree Machine (GOTM) [Zhang B et al., 2004]. The results are shown in Figure 2. Figure 3 shows the distribution of the edge-weight scores generated using these genes on the normal and adenocarcinoma samples from the Harvard dataset. If our method is to be predictive, we expected to see something of a bimodal distribution, although peak height would be dependent on the relative populations of the two groups. This is because weights between members of the same group are expected to be high, while weights between members of different groups are expected to be low. Such a distribution is in fact what we observe in Figure 3.

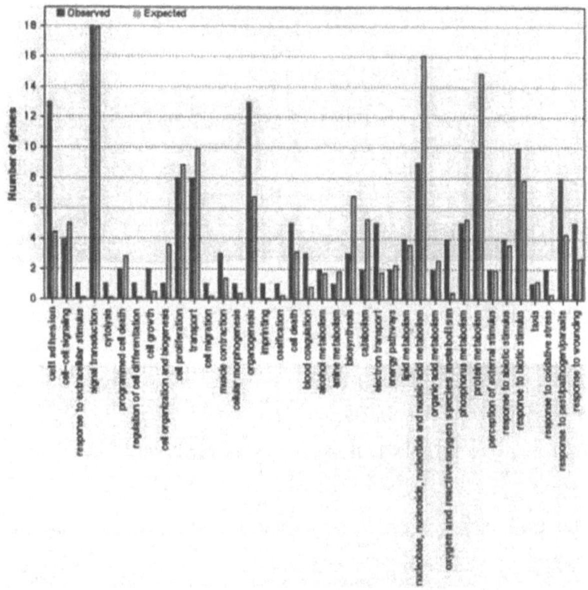

Figure 2. The 84 genes (Michigan data) categorized under gene ontology. Black bars represent observed gene numbers. White bars represent expected gene numbers in the categories. The graph is derived from the fourth annotation level under biological process.

We exploited this property when carrying out threshold selection. We chose an initial threshold slightly to the right of the median edge-weight value. We then enumerated all maximal cliques in the unweighted graph, and checked to see whether every sample is in at least one clique. If not, we chose lower and lower threshold values until we had full coverage (that is, until every sample was in at least one clique). If, on the other hand, our initial threshold gave us full coverage, we incrementally selected higher and higher thresholds until we generated an unweighted graph for which there was at least one sample that was missing from every maximal clique. At this point, we went back one step and used the highest threshold with full coverage. Naturally, this was only one possible method for selecting the threshold; other methods may work equally well. After a suitable threshold was determined, we analyzed the data by testing the supposition that all cliques of significant size were uniform in the sense that they contained samples from adenocarcinoma samples only or from normal samples only.

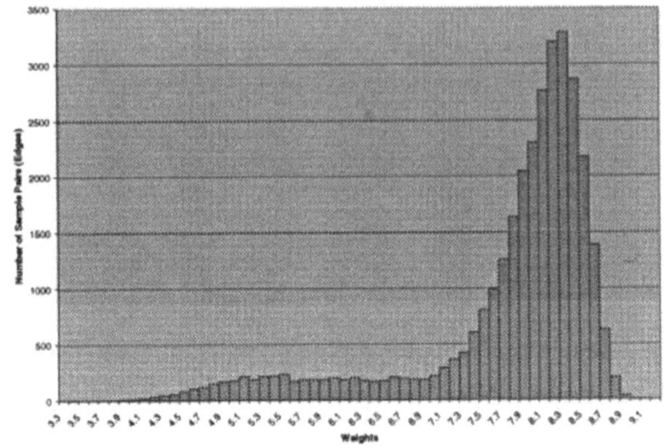

Figure 3. Weights between sample pairs using 78 genes (Harvard data).

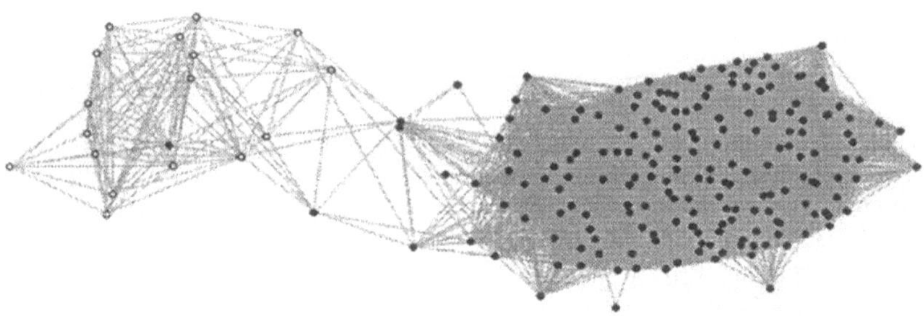

Figure 4. Unweighted graph of the Harvard data set resulting from a threshold of 7.9. Black vertices represent adenocarcinoma samples. White vertices represent normal samples.

When this iterative process was carried out on the Harvard dataset without the use of any previous knowledge pertaining to its sample classifications, we were effectively able to separate the subjects into adenocarcinoma cliques and normal cliques. In fact, at our chosen threshold of 7.9, only one sample out of the 207 combined adenocarcinoma and normal samples was misclassified according to the Harvard dataset using this approach. See Figure 4. This sample is 2001032848AA.CEL. Because it was originally classified as adenocarcinoma but appeared in multiple normal cliques and no adenocarcinoma cliques, we suspected the original classification may have been incorrect. The enumerated cliques histogram is shown in Figure 5. The largest mixed clique was of size six. There were only five mixed cliques in total.

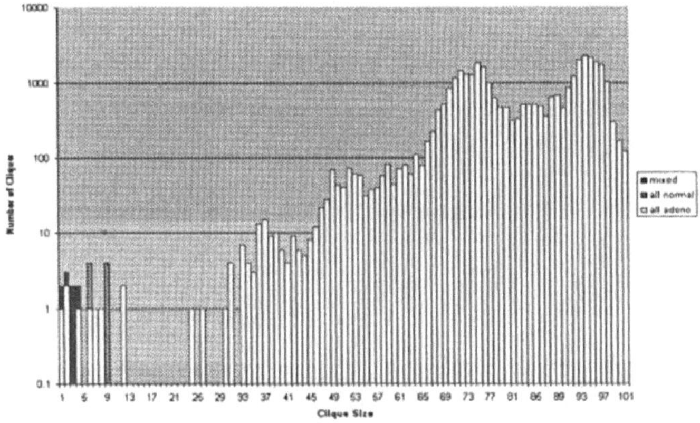

Figure 5. Clique frequency distribution from Harvard data set (adenocarcinoma and normal samples) using 78 genes and a threshold of 7.9.

Of course, we were able to check the quality of our results because the tissue samples represented in the Harvard study were previously classified. To use our methods in the absence of such information, one needs merely to examine the expression values of the highest-scoring genes to determine whether a clique represents a set of adenocarcinoma or normal samples.

5.2 Experiment Two

In this case, we initially identified 195 genes that differentiated cancerous and normal samples. This was reduced to 180 (characterized by gene ontology in Figure 6) using our refinement technique, and 109 of these genes were available in the Michigan dataset.

After following the process we have detailed, we selected a threshold of 8.7. We enumerated maximal cliques on the resulting unweighted graph shown in Figure 7. Our methods were able to sort the samples into cancerous and normal cliques almost flawlessly. In fact, out of the 235 cliques of size three or greater in the resulting graph, only one clique had both cancerous and normal samples, and it was very small (size three). The resultant frequency distribution of these cliques is depicted in Figure 8.

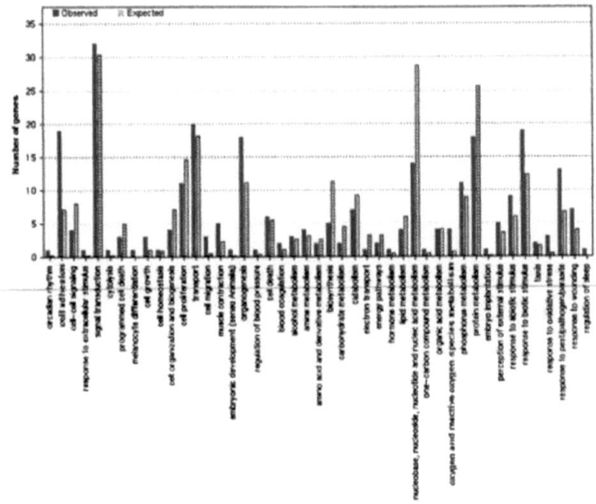

Figure 6. The 180 genes (Harvard data) categorized by gene ontology. Black bars represent observed gene numbers. White bars represent expected gene numbers in the categories.

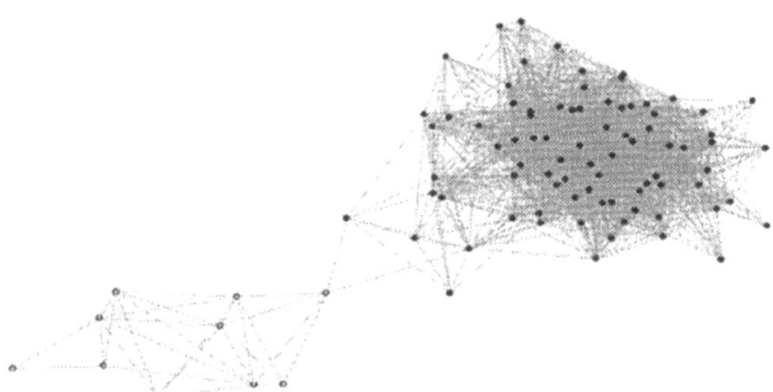

Figure 7. Unweighted graph of the Michigan data set resulting from a threshold of 8.7. Black vertices represent adenocarcinoma samples. White vertices represent normal samples.

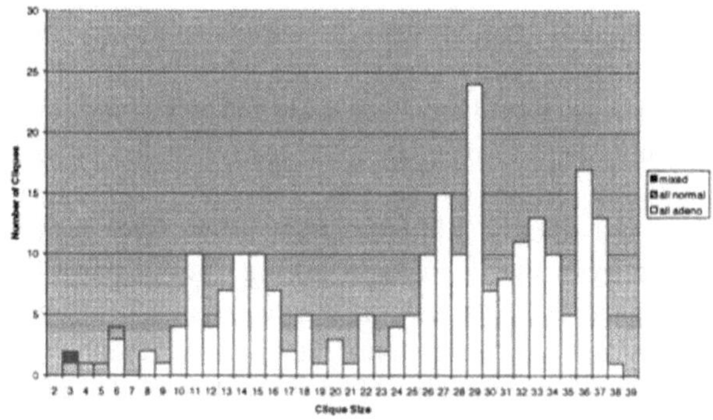

Figure 8. Clique distribution from Michigan data set using 109 genes and a threshold of 8.7.

5.3 Experiment Three

Training on the Harvard dataset to discriminate between adenocarcinoma and squamous cell carcinoma initially gave us 37 genes. After refinement, 35 were left, 26 of which were found in the Stanford data set. In this case, the results given by our method were not as compelling as in the previous two experiments. By using the largest clique containing each sample, we classified 41 out of 47 samples correctly according to the Stanford classifications. Nevertheless, there were still too many mixed cliques. This was not unexpected. Our methods isolated a set of 35 genes as a good discriminator. However, with only 26 of these available in the test dataset, their use provided at best a crude classification tool.

6. CONCLUSIONS

There is no apparent consensus as to the best approach for mining microarray data. Popular methods in current use include Bayesian analysis [Friedman et al., 2000; Sok et al., 2003], hierarchical clustering, and scale-free networks [del Rio et al., 2001], to name just a few. We believe that the novel methodology we have described here can be used to complement these techniques, and also be of independent interest. Deliverables accompanying this effort include the algorithmic framework of our overall strategy, the software tools we have developed and implemented, and of course the resultant gene sets themselves.

A key feature of our approach is the use of two distinct gene-scoring systems, each coupled with a different combinatorial algorithm. One was based on finding optimal cliques within general graphs, the other on isolating near-optimal dominating sets within bipartite graphs. Used in tandem, these algorithms appear to provide an effective means for identifying and ranking predictive genes whose expression levels serve as an accurate discriminator between adenocarcinoma and normal tissues. We emphasize that the use of clique and dominating set together seems to produce better results than would be possible with either approach alone.

The high fidelity with which the resulting cliques partitioned cancerous and normal samples, as illustrated in Figures 6 and 8, prompts us to posit that our methodology has the potential to become the basis for a highly reliable tool for cancer prediction. No *a priori* knowledge of the number of classes contained in the dataset is required. Moreover, it is known that tumor tissue samples are frequently a mixture of multiple types of cells, and that the exact ratio of this mixture is not necessarily consistent, even among samples from the same tumor. Therefore, it is expected that tissue samples might have significant similarity to more than one class, such as adenocarcinoma and normal. This is, in fact, what is observed. Using our method, the classification of the sample is not limited to one class. Nor is the classification based on the highest similarity score. Instead, it is based on the largest (maximal) clique to which the sample belongs. This should result in a higher degree of confidence in our classification.

As a further proof of principle, several of the genes we have identified as discriminators in the Michigan data are known or suspected to play a role in oncogenesis. Among these are: CYP4B1, a cytochrome P450 enzyme that has been implicated in both bladder and lung cancer in humans [Czerwinski et al. 1994; Imaoka et al., 2000]; FHL1, shown to have cytotoxic effects on melanoma cell lines and to possibly play a role in cellular differentiation [de Vries et al., 1975]; the p85 alpha subunit of phosphoinositide-3-kinase, which plays a role in human breast cancer [Das et al., 2003.; Mahabeleshwar et al., 2003]; and tetranectin, which has already been shown to have prognosticative value for survival rates at certain stages of ovarian cancer [Hogdall et al., 2002]. Space limitations prevent us from including a full listing of the genes we have identified here. Thus, we have made this list and additional, extensive details available in the form of a technical report [Langston et al., 2004].

A number of opportunities for future research beckon. For example, the formula we are currently using to assign edge weights relies only on the gene scoring algorithm of our clique-based strategy. This can perhaps be refined by incorporating into it the gene scores computed during our dominating set analysis. Another idea we believe holds promise relies on the use of clique intersection graphs. These are computed as follows. Suppose we are given a filtered, unweighted sample similarity graph, *G*. The vertices of its associated clique intersection graph are the maximal cliques in *G*. Each pair of vertices in the clique intersection graph is connected by an edge if and only if the intersection of the two respective cliques they represent is nonempty. Thus, a clique intersection graph may help to discern the overall structure of relationships contained within sample data. Moreover, cliques within a clique intersection graph may serve to tighten the focus on discriminating factors and act as an aid in quantifying the salient characteristics of archetypical diseased or healthy tissues.

REFERENCES

Abu-Khzam, FN, Collins, RL, Fellows, MR, Langston, MA, Suters, WH, Symons, CT. Kernelization algorithms for the vertex cover problem. *Proceedings, Workshop on Algorithm Engineering and Experiments (ALENEX),* New Orleans, LA, January, 2004.

Abu-Khzam, FN, Langston, MA, Shanbhag, P. Scalable Parallel Algorithms for Difficult Combinatorial Problems: A Case Study in Optimization. *Proceedings, International Conference on Parallel and Distributed Computing and Systems,* Los Angeles, CA, 563-568, November, 2003.

Baldwin, NE, Collins, RL, Langston, MA, Leuze, MR, Symons, CT, Voy, BR. High performance computational tools for motif discovery. *Proceedings, IEEE Workshop on High Performance Computational Biology,* Santa Fe, NM, April, 2004.

Beer, DG, Kardia, SL, Huang, CC, Giordano, TJ, Levin, AM, Misek, DE, Lin, L, Chen, G, Gharib, TG, Thomas, DG, Lizyness, ML, Kuick, R, Hayasaka, S, Taylor, JM, Iannettoni, MD, Orringer, MB, Hanash, S. Gene-expression profiles predict survival of patients with lung adenocarcinoma. *Nature Medicine* 9 (816), 816-824, 2002.

Bhattacharjee, A, Richards, WG, Staunton, J, Li, C, Monti, S, Vasa, P, Ladd, C, Beheshti, J, Bueno, R, Gillette, M, Loda, M, Weber, G, Mark, EJ, Lander, ES, Wong, W, Johnson, BE, Golub, TR, Sugarbaker. DJ, Meyerson, M. Classification of human lung carcinomas by mRNA expression profiling reveals distinct adenocarcinoma subclasses. *Proc Natl Acad Sci USA.* 98 (24), 13790-13795, 2001.

Czerwinski, M, McLemore, TL, Gelboin, HV, Gonzalez, FJ. Quantification of CYP2B7, CYP4B1, and CYPOR messenger RNAs in normal human lung and lung tumors. *Cancer Res.* 54(4): 1085-91, 1994.

Das, R, Mahabeleshwar, GH, Kundu, GC. Osteopontin stimulates cell motility and nuclear factor kappaB-mediated secretion of urokinase type plasminogen activator through phosphatidylinositol 3-kinase/Akt signaling pathways in breast cancer cells. *J Biol Chem.* 278(31):28593-606, 2003.

R. G. Downey and M. R. Fellows. *Parameterized Complexity.* Springer-Verlag. 1999.

Friedman, N, Linial, M, Nachman, I, Pe'er, D. Using Bayesian networks to analyze expression data. *J Comput Biol.* 7(3-4):601-20, 2000.

Garber, ME, Troyanskaya, OG, Schluens, K, Petersen, S, Thaesler, Z, Pacyna-Gengelbach, M, van de Rijn, M, Rosen, GD, Perou, CM, Whyte, RI, Altman, RB, Brown, PO, Botstein, D, Petersen, I. Diversity of gene expression in adenocarcinoma of the lung. *Proc Natl Acad Sci U S A.* 98(24): 13784-13789, 2001.

Garey, MR, Johnson, DS. *Computers and Intractability.* W. H. Freeman, New York, 1979.

Hogdall, CK, Norgaard-Pedersen, B, Mogensen, O.The prognostic value of pre-operative serum tetranectin, CA-125 and a combined index in women with primary ovarian cancer. *Anticancer Res.* 22(3): 1765-8, 2002.

Hu, JH, Yin, GS, Morris, JS, Zhang, L, Wright, FA. Entropy and survival-based weights to combine Affymetrix array types in the analysis of differential expression and survival. *Critical Assessment of Microarray Data Analysis "CAMDA'03": Oral and Poster Presenters Abstracts,* 78-82, 2003.

Imaoka, S, Yoneda, Y, Sugimoto, T, Hiroi, T, Yamamoto, K, Nakatani, T, Funae, Y. CYP4B1 is a possible risk factor for bladder cancer in humans. *Biochem Biophys Res Commun.* 277(3):776-80, 2000.

Irizarry, RA, Hobbs, B, Collin, F, Beazer-Barclay, YD, Antonellis, KJ, Scherf, U, Speed, TP. Exploration, normalization, and summaries of high density oligonucleotide array probe level data. *Biostatistics* 4(2): 249-264. 2003.

Langston, MA, Lin, L, Peng, X, Baldwin, NE, Symons, CT, Zhang, B, Snoddy, JR. A combinatorial approach to the analysis of differential gene expression data. Technical Report UT-CS-04-514, Dept. of Computer Science, University of Tennessee, 2004.

Mahabeleshwar, GH, Kundu, GC. Syk, a protein-tyrosine kinase, suppresses the cell motility and nuclear factor kappa B-mediated secretion of urokinase type plasminogen activator by inhibiting the phosphatidylinositol 3'-kinase activity in breast cancer cells. *J Biol Chem.* 278(8):6209-21, 2003.

del Rio, G, Bartley, TF, del-Rio, H, Rao, R, Jin, KL, Greenberg, DA, Eshoo, M, Bredesen, DE. Mining DNA microarray data using a novel approach based on graph theory. *FEBS Letters* 509(2):230-4, 2001.

Sok, JC, Kuriakose, MA, Mahajan, VB, Pearlman, AN, DeLacure, MD, Chen, FA. Tissue-specific gene expression of head and neck squamous cell carcinoma in vivo by complementary DNA microarray analysis. *Arch Otolaryngol Head Neck Surgery* 129(7):760-70, 2003.

de Vries, JE, Meyering, M, van Dongen, A, Rumke, P. The influence of different isolation procedures and the use of target cells from melanoma cell lines and short-term cultures on the non-specific cytotoxic effects of lymphocytes from healthy donors. *Int J Cancer.* 15(3): 391-400, 1975.

Zhang, B, Schmoyer, D, Kirov, S, Snoddy, J. GOTree Machine (GOTM): a web-based platform for interpreting sets of interesting genes using gene ontology hierarchies. To appear in *BMC Bioinformatics,* 2004; http://genereg.ornl.gov/gotm.

Chapter 17

GENES ASSOCIATED WITH PROGNOSIS IN ADENOCARCINOMA ACROSS STUDIES AT MULTIPLE INSTITUTIONS

Andrew V. Kossenkov[1,2], Ghislain Bidaut[1,3], Michael F. Ochs[1,*]
[1]*Bioinformatics, Fox Chase Cancer Center, Philadelphia, PA;* [2]*Drexel University, Philadelphia, PA;* [3] *Structural and Genetic Information Lab, CNRS-AVENTIS. Marseille, France*

Abstract: Cancer is a complex disease, comprising many different specific malfunctions within the body. Because many biological processes occur simultaneously within all cells, the gene expression related to tumor behavior is generally confounded with expression due to routine metabolic processes and additional processes unrelated to tumorigenesis. Bayesian Decomposition has been used to isolate expression signatures related to these processes as well as signatures related to patient prognosis. The signatures related to prognosis have been analyzed to identify biological processes as well as specific genes whose presence appears related to outcome in all studies.

Key words: Bayesian methods, gene expression, gene ontology, cancer

1. INTRODUCTION

Although treatment regimens undergo constant improvement, cancer remains the second leading cause of death throughout the Western world [Alison et al., 1997]. Targeted treatment and individualized medicine offer hope for improved outcomes. However, a deep understanding of cancer development in individual malignancies is required. This demands information on the process that led to the specific cellular malfunction present in the cancer cells. Since the development of cancer generally

[*] author to whom correspondence should be addressed

involves the cellular signaling networks that control cell growth, differentiation, apoptosis, and motility [Kolch, 2000; Jacks et al., 2002], the extreme complexity of these pathways and the multiple failure points and checkpoints lead to the reality that observed cancers arise from a myriad of different cellular malfunctions [Cooper, 1992; Macdonald et al., 1997]. It is from this complex background that microarray analysis attempts to glean insight to improve cancer treatment.

Identifying cellular malfunctions in cancer at early stages remains a critical issue for improving patient survival. The studies in the CAMDA 2003 data set are primarily focused on refinement of the identification of the type of cancer using computational and statistical approaches, as was the focus of a number of early studies using microarrays [Golub et al., 1999; Alizadeh et al., 2000; Zhang et al., 2001]. These methods can also be extended to the discovery of biomarkers in the form of differential levels of production of mRNA [Carr et al., 2003; Kikuchi et al., 2003; Williams et al., 2003], which has the advantage of providing a more viable clinical protocol. These methods generally apply microarray technology to detect disease state from tissue samples, aiming to refine identification of suspect tissues after a biopsy has been performed. The additional information can aid in tailoring treatment, as histologically different cancers require substantially different therapeutic regimens to maximize patient survival.

While the techniques noted above are useful, they have certain limitations as regards more advanced uses in cancer research. Cancer is primarily a disease of signaling, and newer therapeutics specifically target proteins involved in cellular signaling [Mauro et al., 2001; Repka et al., 2003; von Mehren, 2003]. It is therefore highly desirable to understand the key genes that have triggered tumorigenesis, especially in cases where these genes might be shared across multiple individuals, providing a useful target for therapeutic development. Genes playing a role in prognosis in multiple individuals are also more likely to encode proteins serving as triggers to cancer development than genes not shared, if indeed specific cancers, such as adenocarcinoma, arise from common events. Microarray measurements are being used to provide insight into these questions. Here we focus on the use of Bayesian Decomposition [Bidaut et al., 2002; Moloshok et al., 2002; Moloshok et al., 2003], together with construction of relational trees between multiple analyses and gene ontology information, in order to understand the processes at work and the pivotal genes triggering the development of adenocarcinoma and poor prognosis.

2. METHODS

2.1 Data Processing

The initial data were downloaded from the CAMDA web site (http://www.camda.duke.edu/camda03/). These data included Affymetrix CEL files for Harvard and Michigan data sets and files in GenePix format for the Stanford data set. The Ontario data was also downloaded but was not used in the analysis due to a lack of overlap in annotated genes with the other sets. Only adenocarcinoma and normal samples were used in the analysis. CEL files of Harvard and Michigan data sets were processed with dChip software, and expression levels were calculated using the PM/MM, i.e. perfect match to mismatch, difference model with the median overall intensity array method defining the baseline [Cheng et al., 2001]. The standard error across a probe set was used as the uncertainty of the measurement of each expression level in Bayesian Decomposition. For the Stanford data set, the mean background and foreground intensities of channel 1 and channel 2 from the GenePix measurements were used. Normalization was performed with the Functional Genomics Data Pipeline [Grant et al., 2004], using LOESS with a smoothing parameter of 0.9 to equalize the channels, and expression ratios were calculated. Due to the lack of replicates in this data, the uncertainty of each measurement had to be estimated and was set to 30% of the expression level based on previous experience. For those data points where the ratio was negative or data was missing, the ratio was set to 1.0 and the uncertainty to 289 (equal to maximum ratio across all data points), effectively insuring that these data points would not affect the model.

2.2 Data Annotation

The genes present in the Harvard, Michigan and Stanford data sets were annotated for gene ontology information [Ashburner et al., 2000] using the Automated Sequence Annotation Pipeline [Kossenkov et al., 2003] as depicted in Figure 1. Since the goal was to identify genes consistently linked to outcome across all studies, only genes with annotations in all three data sets were retained for analysis. As the analysis also focused on differences in expression between tissue types, the coefficient of variation was calculated for each transcript in all three data sets (expression levels for Affymetrix data, expression ratios for spotted array data), and only genes for which these exceeded an arbitrary cutoff of 35% were retained. The final data comprised 1216 transcripts from the Harvard data, 1088 from the

Michigan data, and 1337 transcripts from the Stanford data, representing 987 unique Unigene clusters.

The samples from the three data sets were classified by tumor stage according to the information provided. The Harvard data set comprised four classes (76 first stage, 24 second stage, 13 third and fourth stage, 17 normal), the Michigan data set comprised three classes (67 first stage, 19 third stage, 10 normal) and the Stanford data set comprised three classes (17 second stage, 15 third stage, five normal).

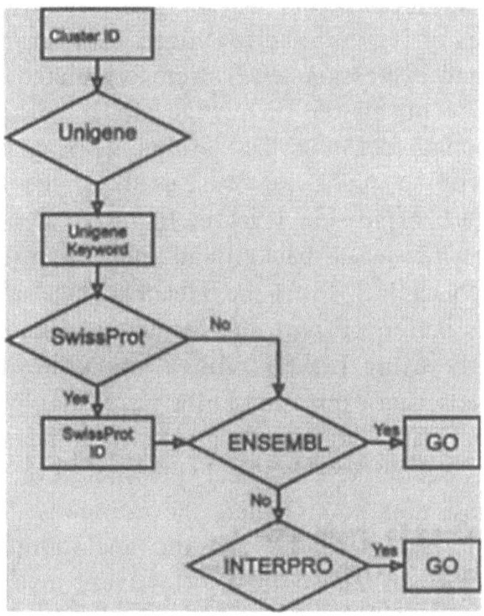

Figure 1. The annotation method used to annotate genes and determine the Gene Ontology information. For each sequence spotted on the array, updated Unigene information was retrieved. The Unigene Keyword was used to search the Swiss-Prot database. If a match was found the Swiss-Prot ID was used to retrieve gene ontology information for the clone from Ensembl or Interpro databases.

2.3 Bayesian Decomposition

Bayesian Decomposition was used to analyze these data sets separately, exploring different potential numbers of patterns. Bayesian Decomposition is a matrix decomposition algorithm that allows the encoding of additional prior information within a Bayesian framework [Besag et al., 1995]. The input is a set of data in the form of a matrix, $\boldsymbol{D}$, which describes the measurements of expression levels for genes (rows) across different tissues (columns). In addition, there is a matrix $\boldsymbol{\varepsilon}$ that provides estimates of the uncertainty or noise for each individual measurement in ***D***. From these data, two matrices are constructed such that

$$\mathbf{D} = \mathbf{AP} + \boldsymbol{\varepsilon} \tag{1}$$

where ***P*** contains *k* rows giving *k* patterns within the data across the tissues, and ***A*** provides a measure of how strongly each gene contains each pattern. The Markov chain sampling provides both mean value and standard deviation (σ) estimates for each matrix element in ***A*** and ***P***. The rows of ***P*** are normalized to sum to one, in order to resolve the inherent scale invariance.

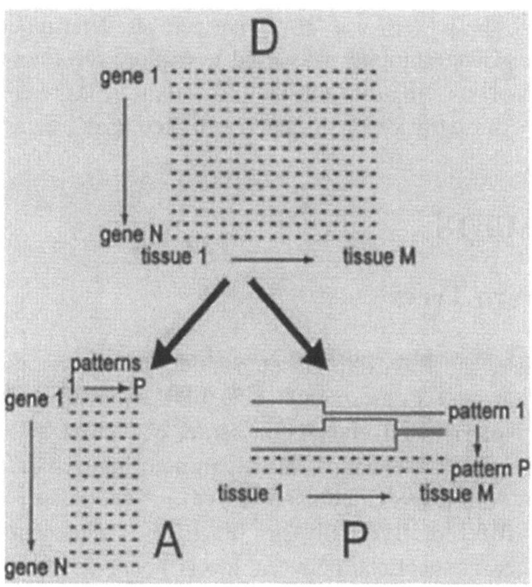

Figure 2. The decomposition performed by Bayesian Decomposition on the data. The tissues are separate samples and the patterns explain variations across the samples. The first three patterns depicted here are enforced to be related to the tumor staging or type, but this number will vary depending on the data set.

Two separate full analyses were preformed. In the first, the tumor staging was included by enforcing the existence of patterns related to each stage. In the second, there were no enforced patterns, so that staging information was ignored. For both cases, the Harvard data were analyzed positing from four to 13 patterns, while the Michigan and Stanford data were analyzed positing from three to 12 patterns. The number of patterns was chosen to provide between zero and nine additional patterns to explain the non-tumor stage related behaviors (e.g., routine metabolism) present in the data, as demonstrated in model organisms previously [Moloshok et al., 2002]. The use of additional patterns provides freedom for outcome-related patterns not determined by tumor staging to emerge.

The results were analyzed across different numbers of posited patterns independently for each data set to identify stable patterns linked to prognosis. Pearson correlation coefficients between outcome and pattern amplitudes for each pattern were calculated, and genes linked to these patterns were identified. The patterns were considered robust if they showed correlation across different numbers of posited patterns. A gene was considered associated with a pattern if its amplitude in the column of the ***A*** matrix linked to the pattern was 3σ above zero as determined by Bayesian Decomposition. Gene ontology was used to explore the roles of the patterns linked with prognosis. In addition, the individual genes were then compared to identify only those present in all patterns linked with prognosis.

3. RESULTS

3.1 Pattern Trees

The analysis of the Bayesian Decomposition output was performed using ClutrFree [Bidaut and Ochs, 2004], a visualization tool allowing global comparison of patterns and clusters in terms of shapes and robustness of gene assignment. For the Harvard, Michigan and Stanford data, pattern trees were created by ClutrFree (Figures 3 and 4). The pattern tree represents links between different BD analyses. Each level represents a single BD analysis, with the connections between levels being determined by Pearson correlation values between individual patterns. The links are created in a greedy way, with the overall highest correlation connected first, then the highest for remaining patterns, etc. The thickness of the line connecting nodes (i.e., patterns) shows the strength of the correlation. In Figure 3, the first few patterns in each run are locked to tumor staging in a way analogous to Figure 2. For instance, for the Michigan data (Figure 3a), the first group (all 1 in nodes in Figure 3) is stage 1 tumor, the second group (2) is stage 3 tumor, and the third group (3) is normal tissue. Additional patterns are free to contain any of the samples at any strength (i.e., amplitude).

In order to find the gene expression patterns related to survival data, Pearson correlation coefficients between outcome and pattern amplitudes for each pattern of each Bayesian Decomposition run were calculated. For example, Table 1 provides all correlation coefficients for patterns in all analyses of the Michigan data set. Then the pattern trees were examined to find patterns with largest outcome correlations that appeared in one persistent branch. This is shown in Figure 3a for patterns from the Michigan data, where highlights indicate pattern 5 from 9 posited patterns, pattern 8 from 10 posited patterns, pattern 11 from 11 posited patterns and pattern 9 from 12 posited patterns.

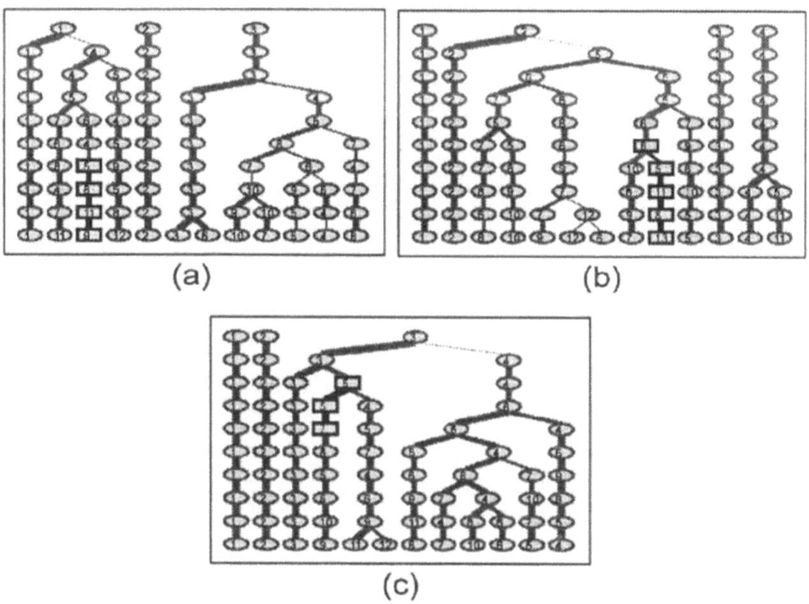

Figure 3. Trees relating the multiple analyses performed by Bayesian Decomposition including tumor staging. Results for Michigan are shown in (a), Harvard in (b), and Stanford in (c), with all patterns correlated with survival and persistent in the tree highlighted with black rectangles. Comparison with Table 1 shows that in (a), pattern 6 of 7, two above the uppermost black rectangle, is also correlated with outcome. However, because pattern 4 of 8 is not, pattern 6 of 7 is excluded for not being persistent. The numbers are for housekeeping purposes, allowing columns of the ***A*** matrix to be linked to rows of the ***P*** matrix.

Thus, for each data set persistent patterns within branches were identified. Persistent patterns comprise a linked series of nodes that all correlate with outcome. Five such patterns were found in the Harvard data, four patterns in the Michigan data, and three patterns in the Stanford data (as denoted by the black rectangles in Figure 3). Lists of genes linked to these expression patterns were then generated for each of the patterns, and the gene ontologies for these genes were analyzed.

Table 1. Correlation coefficients between outcome and pattern amplitudes for the Michigan data. Coefficients greater than 0.6 are in bold. Rows are the number of patterns posited in a run (corresponding to rows in the tree in Figure 3a), columns are the pattern ordinal numbers in the run (corresponds to the number in an oval from Figure 3).

	4	5	6	7	8	9	10	11	12
4	0.35								
5	0.44	0.00							
6	0.46	0.39	-0.12						
7	-0.23	0.42	**0.65**	0.22					
8	0.43	-0.25	0.47	0.49	-0.08				
9	0.27	**0.66**	0.13	0.13	0.34	-0.23			
10	0.44	-0.19	0.05	-0.06	**0.74**	-0.01	0.16		
11	-0.12	-0.17	0.07	0.30	-0.15	0.10	0.43	**0.62**	
12	0.13	0.01	-0.20	0.34	0.09	**0.72**	0.02	0.39	-0.23

The same procedure was performed for the Bayesian Decomposition analyses that did not include staging information. The results of these analyses are shown in Figure 4 in a manner matching that used in Figure 3.

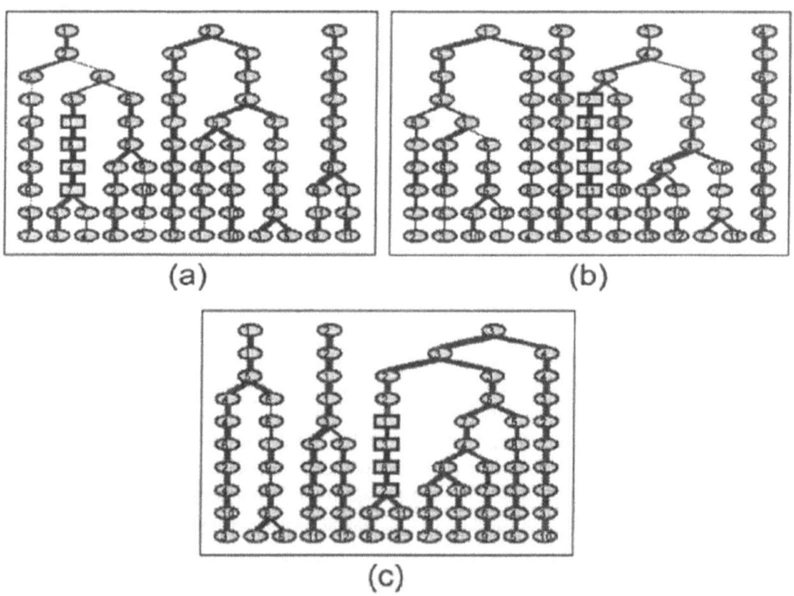

Figure 4. Trees relating the multiple analyses performed by Bayesian Decomposition excluding tumor staging. Results for Michigan are shown in (a), Harvard in (b), and Stanford in (c), with all patterns correlated with survival and persistent in the tree highlighted with black rectangles.

3.2 Analysis of Gene Ontology

For persistent patterns with significant correlation with outcome, we explored enhancements in terms of gene ontology using ClutrFree. Enhancement is computed as a ratio of the number of genes with a gene ontology term in a pattern normalized by the total number of genes in the pattern and the number of genes with the gene ontology term in the full data set normalized by the total number of genes studied.

A gene ontology term was considered as 'occurred' in a pattern if the enhancement was greater than 1 in at least 10 of the 12 patterns linked to prognosis and as 'removed' if the enhancement was less than 1 in at least 10 of the 12 patterns. Table 2 comprises a portion of that gene ontology list. The list shows that the patterns linked to prognosis include terms related to development (tumors often have activation of normally silent developmental genes), cell migration, signaling activity, and RAS and EGF activity (both linked to tumorigenesis). Terms that have low representation include the regulation of cell proliferation, negative regulation of NF-kappaB, and inactivation of MAPK activity, all of which can be considered related to control of unconstrained cell growth and oncogenesis.

Table 2. 'Occurred' and 'removed' Gene Ontologies. The number of patterns where the GO term occurred or was removed is shown in parantheses.

Occurred	Removed
GO:0008406 gonad development (12)	GO:0042127 regulation of cell proliferation (12)
GO:0016477 cell migration (11)	GO:0008543 FGF receptor signaling pathway (11)
GO:0001701 embryonic development (11)	GO:0008283 cell proliferation (11)
GO:0000187 activation of MAPK (11)	GO:0007253 cytoplasmic sequestering of NF-kappaB (11)
GO:0016055 Wnt receptor signaling pathway (10)	GO:0042347 negative regulation of NF-kappaB (10)
GO:0007265 RAS protein signal transduction (10)	GO:0007261 STAT protein dimerizaton (10)
GO:0007186 G-protein coupled receptor protein signaling pathway (10)	GO:0007229 integrin mediated signaling pathway (10)
GO:0007173 EGF receptor signaling pathway (10)	GO:0000188 inactivation of MAPK (10)
	GO:0000074 regulation of cell cycle (10)

3.3 Genes Associated with Prognosis

From the 12 patterns linked to prognosis in the analysis including staging information, consistent genes were identified. First, genes that were robust in assignment along the highlighted branches in Figure 3 were identified. The sets contained 238 genes for Michigan, 397 genes for Harvard, and 366 genes for Stanford. These lists were then compared to identify those genes linked to prognosis in all studies.

The intersection of the three studies yielded 45 genes that were suspected to be associated with survival in adenocarcinoma, of which 27 had annotations, as shown in Table 3. As can be seen, most of these genes have previously been shown to be involved in cancer. A number of them are known to specifically modulate cell motility (EDG2, LAMA3, ADD3, CD58, SELE, PTPRR), a key issue in metastasis, which naturally has a major impact on prognosis. Others have been implicated in apoptosis (VDR) and prognosis in various cancers (IL15, SULT1C1, PTHLH). PTHLH is of special interest as it has been linked to tumor progression in lung cancer. Another gene of interest is XRCC4 that encodes a double strand break repair enzyme. Such enzymes serve with important checkpoint control proteins to guarantee that cells with substantial DNA damage do not continue division.

3.4 Results without Inclusion of Staging Information

As noted above, the Harvard, Michigan and Stanford data were also analyzed without using information about tumor stage. Figure 4 shows the results, with one branch in each case being linked to outcome just as in the analysis using information on staging. For the Affymetrix data sets (i.e., Michigan and Harvard), the patterns linked to prognosis appear at fewer total patterns, perhaps because they can emerge earlier without the constraints.

For Michigan, 330 genes were consistent in these patterns, with 202 of these in common with the 238 genes identified when staging information was included. For Harvard 399 genes were consistent in these patterns, with 323 genes in common with the 397 genes identified when staging information was included. However, for Stanford only 22 genes were consistent in these patterns, with only eight in common with the 366 genes identified with staging information included.

Because of the small number of genes from Stanford, the intersection of all three studies included only two genes, QDPR (quinoid dihydropteridine reductase) and TCF7 (transcription factor 7, T-cell specific). Only QDPR was in the original list.

Table 3. Genes identified by intersection of the 12 gene lists. There are six immune system genes, 18 genes with links or potential links to cancer, and six genes directly linked to cell adhesion or metastasis.

GENE	IMMUNE RESPONSE	KNOWN GENE FUNCTIONS
EDG2		lysophosphatidic acid (LPA) receptor, link to G protein and activation of signaling cascade, essential for normal development in mice, modulate cell motility under pathological conditions
SURB7		
LAMA3		laminin 5 gene, essential in stability of cutaneous basement membrane zone
QDPR		
IL15RA	IL15 receptor	binding activates JAK-STAT pathway active in many cancers
IL15	IL15	post translational modifications critical, errors in IL-15 control in mice lead to lethal leukemia
TDO2		possible candidate in serotonin metabolism difficulties with link to alcoholism, ADHD, Tourette's syndrome
NCF4		
ADD3		adducin, component of actin cytoskeletal cortex, stabilized integral membrane proteins
MBL2	innate immunity	mannose binding lectin,
SULT1C 1		converts estrogens to sulfated forms, key role in breast cancer due to effect on hormones and response to tamoxifen
PPID		
CNR1		brain receptor protein, suspected to dictate formation of synaptic connections in the brain
DPYD		variant forms cannot metabolize 5-FU, critical role in breaking down 5-FU
XRCC4		double stranded break repair enzyme
CD58	governs cell-cell binding	may play a role in cell-cell adhesion
ITGAL		
SELE		E selectin, specific adhesive molecule for cell-vascular endothelial cells, linked to metastasis
CTH		
PTHLH		significantly linked to tumor progression in lung cancer
WNT5A		signaling protein involved in development, WNT signaling linked to tumor formation when aberrantly activated
IRF7	IFN regulatory factor	linked to activation of numerous IFN-alpha family members including possible NF-kappaB and c-jun
BTN3A1		
IL1R1	IL1 receptor	linked to rheumatoid arthritis, potent activator of transcription
GALNT3		
VDR		vdr signaling linked to growth arrest, differentiation, and induction of apoptosis in breast cancer
PTPRR		tyrosine kinase receptors linked to induction of signaling cascades linked to cell division, migration, and survival

Focusing only on the Harvard and Michigan results, 21 genes are in common between this analysis ignoring staging and the analysis including all three data sets with staging information. These are SELE, QDPR, CNR1, ADD3, BTN3A1, MBL2, CTH, PTPRR, NCF4, TDO2, IL15, DPYD, WNT5A, and IL15RA.

4. DISCUSSION

Adenocarcinoma of the lung continues to take a heavy toll in mortality despite improved diagnostic techniques and treatments. New technologies, such as microarrays and proteomics, can potentially enhance our understanding of this disease and aid in diagnosis and treatment planning. Many applications of these new technologies aim to refine diagnosis through identification of patterns or signatures linked to specific phenotype, such as response to therapy. While these techniques can be valuable, they have serious limitations, especially in microarray applications. Microarray techniques for tumors are presently invasive, requiring at minimum a biopsy sample and more typically a tumor mass. As such, it must be the case that cancer is already at minimum suspected, a mass has been identified, and a biopsy obtained. Therefore for diagnostic approaches, serum proteomics is a far more promising domain, since it provides a minimally invasive procedure and the potential for identification of signatures of early tumorigenesis [Petricoin et al., 2002].

However, in other regards microarrays provide far more detailed insights into cellular state than proteomics. With a microarray the full genome can now routinely be explored in terms of production of mRNA, increasingly including splice variants. Proteomics remains a field where only high abundance proteins (e.g., in 2D gels) or highly studied proteins (e.g., those with monoclonal antibodies) can be explored presently. The deep view of the cellular machinery provided in microarrays offers the opportunity to identify cellular processes through the linking of changes to activity of signaling pathways, gene ontologies, and potentially gene networks. As such, microarrays provide a global view of how the cellular machinery responds in different tumor cells, what genes appear significantly changed in these cases, and therefore they can provide information on potential targets for new therapeutics.

Here we have relied on the assumption that adenocarcinoma is a disease whose key underlying physiology is independent of the specifics of the center treating the patient. We have then applied Bayesian Decomposition to identify groups of genes linked together in patterns, relying on its ability to handle multiple groupings, since genes encode proteins that serve multiple purposes in cells. These patterns have been correlated with survival, which is possible as each pattern also has a strength of association with each patient. Patterns which appear to stably relate to prognosis as the number of potential patterns increase were then chosen for exploration of gene membership. We have used gene ontology information to explore the potential biological purpose of the patterns associated with prognosis. Gene ontology indicates losses in processes regulating cellular proliferation and increases in developmental processes and some key signaling processes. In addition, a group of 47 genes having a strong link to prognosis in all centers was identified, providing a list for validation and potential followup as therapeutic targets. Known genes in this list include a number of known cancer- and prognosis-related proteins.

We also applied Bayesian Decomposition in a way that did not include known pathology. Here the results were minorly different for the centers relying on Affymetrix technology, however they changed rather dramatically for the one study using spotted arrays. It is possible that this reflects an inherent problem with the use of nonreplicated spotted arrays, since we have little information on the uncertainty of individual gene expression measurements and rely on a global uncertainty estimate. However, since it is also the smallest study in this analysis, it is possible that the inclusion of tumor staging information provides added statistical power. Since staging effectively links individual samples, patterns with these enforced links effectively average over samples during analysis. This may provide enough additional power to provide better insight.

The work presented here represents an approach to microarray analysis that stresses exploration of potential biologically significant areas determining phenotype. The next step with such an approach is validation of key genes by real-time PCR analysis and generation of hypotheses for testing in cell lines and model organisms.

5. ACKNOWLEDGEMENTS

We thank the National Institutes of Health, National Cancer Institute (CCCG CA06927 to R. Young), the Pennsylvania Department of Health (grant to M. F. Ochs), and the Pew Foundation for support.

6. REFERENCES

Alison, M and Sarraf, C (1997) Understanding Cancer. Cambridge University Press, Cambridge.

Alizadeh, AA, Eisen, MB, Davis, RE, Ma, C, Lossos, IS, Rosenwald, A, Boldrick, JC, Sabet, H, Tran, T, Yu, X, Powell, JI, Yang, L, Marti, GE, Moore, T, Hudson, J, Jr., Lu, L, Lewis, DB, Tibshirani, R, Sherlock, G, Chan, WC, Greiner, TC, Weisenburger, DD, Armitage, JO, Warnke, Levy R, Wilson W, Grever MR, Byrd JC, Botstein D, Brown PO, R, Staudt, LM (2000) Distinct types of diffuse large B-cell lymphoma identified by gene expression profiling. Nature 403: 503-11.

Ashburner, M, Ball, CA, Blake, JA, Botstein, D, Butler, H, Cherry, JM, Davis, AP, Dolinski, K, Dwight, SS, Eppig, JT, Harris, MA, Hill, DP, Issel-Tarver, L, Kasarskis, A, Lewis, S, Matese, JC, Richardson, JE, Ringwald, M, Rubin, GM and Sherlock, G (2000) Gene ontology: tool for the unification of biology. The Gene Ontology Consortium. Nat Genet 25: 25-9.

Besag, J, Green, P, Higdon, D and Mengersen, K (1995) Bayesian computation and stochastic systems. Statistical Science 10: 3 - 66.

Bidaut, G, Moloshok, TD, Grant, JD, Manion, FJ and Ochs, MF (2002). Bayesian Decomposition analysis of gene expression in yeast deletion mutants. Methods of Microarray Data Analysis II. Johnson, K and Lin, S. Boston, Kluwer Academic: 105-122.

Bidaut, G and Ochs, MF (In Press) ClutrFree: Cluster tree visualization and interpretation. Bioinformatics.

Carr, KM, Bittner, M and Trent, JM (2003) Gene-expression profiling in human cutaneous melanoma. Oncogene 22: 3076-80.

Cheng, L and Wong, WH (2001) Model-based analysis of oligonucleotide arrays: Expression index computation and outlier detection. Proc Natl Acad Sci U S A 98: 31 - 36.

Cooper, GM (1992) Elements of Human Cancer. Jones and Bartlett Publishers, Boston.

Golub, TR, Slonim, DK, Tamayo, P, Huard, C, Gaasenbeek, M, Mesirov, JP, Coller, H, Loh, ML, Downing, JR, Caligiuri, MA, Bloomfield, CD and Lander, ES (1999) Molecular classification of cancer: class discovery and class prediction by gene expression monitoring. Science 286: 531-7.

Grant, JD, Somers, LA, Zhang, Y, Manion, FJ, Bidaut, G and Ochs, MF (2004) FGDP: functional genomics data pipeline for automated, multiple microarray data analyses. Bioinformatics 20: 282 - 283.

Jacks, T and Weinberg, RA (2002) Taking the study of cancer cell survival to a new dimension. Cell 111:923-5.

Kikuchi, T, Daigo, Y, Katagiri, T, Tsunoda, T, Okada, K, Kakiuchi, S, Zembutsu, H, Furukawa, Y, Kawamura, M, Kobayashi, K, Imai, K and Nakamura, Y (2003) Expression profiles of non-small cell lung cancers on cDNA microarrays: identification of genes for prediction of lymph-node metastasis and sensitivity to anti-cancer drugs. Oncogene 22: 2192-205.

Kolch, W (2000) Meaningful relationships: the regulation of the Ras/Raf/MEK/ERK pathway by protein interactions. Biochem J 351 Pt 2: 289-305.

Kossenkov, A, Manion, FJ, Korotkov, E, Moloshok, TD and Ochs, MF (2003) ASAP: automated sequence annotation pipeline for web-based updating of sequence information with a local dynamic database. Bioinformatics 19: 675-676.

Macdonald, F and Ford, CHJ (1997) Molecular Biology of Cancer. BIOS Scientific Publishers, Ltd., Oxford.

Mauro, MJ and Druker, BJ (2001) STI571: targeting BCR-ABL as therapy for CML. Oncologist 6: 233-8.

Moloshok, TD, Datta, D, Kossenkov, AV and Ochs, MF (2003). Bayesian Decomposition classification of the Project Normal data set. Methods of Microarray Data Analysis III. Johnson, KF and LIn, SM. Boston, Kluwer Academic: 211 - 232.

Moloshok, TD, Klevecz, RR, Grant, JD, Manion, FJ, Speier, WFt and Ochs, MF (2002) Application of Bayesian Decomposition for analysing microarray data. Bioinformatics 18: 566-75.

Petricoin, EF, Ardekani, AM, Hitt, BA, Levine, PJ, Fusaro, VA, Steinberg, SM, Mills, GB, Simone, C, Fishman, DA, Kohn, EC and Liotta, LA (2002) Use of proteomic patterns in serum to identify ovarian cancer. Lancet 359: 572-7.

Repka, T, Chiorean, EG, Gay, J, Herwig, KE, Kohl, VK, Yee, D and Miller, JS (2003) Trastuzumab and interleukin-2 in HER2-positive metastatic breast cancer: a pilot study. Clin Cancer Res 9: 2440-6.

von Mehren, M (2003) Recent advances in the management of gastrointestinal stromal tumors. Curr Oncol Rep 5: 288-94.

Williams, NS, Gaynor, RB, Scoggin, S, Verma, U, Gokaslan, T, Simmang, C, Fleming, J, Tavana, D, Frenkel, E and Becerra, C (2003) Identification and validation of genes involved in the pathogenesis of colorectal cancer using cDNA microarrays and RNA interference. Clin Cancer Res 9: 931-46.

Zhang, H, Yu, CY, Singer, B and Xiong, M (2001) Recursive partitioning for tumor classification with gene expression microarray data. Proc Natl Acad Sci U S A 98: 6730-5.

This page intentionally left blank

Index

Adenocarcinoma, 147-148, 157-158,165, 250
Bayesian methods, 207, 239
C5.0, 151-153
CART, 152
Censoring, 23, 109, 115, 116
Classification, 2, 3, 6, 18, 32, 33, 65, 80, 89, 94, 108, 130, 144, 160-162, 167, 173, 190, 204, 218, 222, 237
  Neural network classifiers, 211
Combinatorial methods, 223
Conditional probability, 140
Cox proportional hazards model, 28-29, 33, 51, 54, 56, 60, 63-64, 67, 69-70, 72-76, 78-79, 91, 95, 103-105, 110, 163, 165, 167-171, 189
Decision trees, 147, 151
Differential correlation, 121, 129
Differentially expressed genes, 35
EMMIX-GENE algorithm, 163-164
Entropy, 108, 238
Features, 135
Gene ontology, 78, 244, 251-252
Gene selection, 173, 204, 208, 210
Information gain, 152
Integration, 81, 84, 123, 147, 149
Interaction, 90
Kaplan-Meier, 21, 67, 150, 154-155, 163, 165, 168-171, 176

k-NN, 151-153

Leukemia, 148, 159
Lung cancer, 1, 9-11, 14, 19, 147-148, 157, 159 163
  Lung adenocarcinoma, 121, 161, 205
  Non-small cell lung cancer (NSCLC), 18, 51, 62, 65
Meta-analysis, 33, 67
MLPs, 151
Multilayer perceptrons, 151
Oligonucleotide microarrays
  Affymetrix, 22, 25, 33, 51, 53-58, 64, 67-68, 81-84, 94-95, 98-99, 123, 134, 148, 150, 164, 167, 175, 207, 209, 214, 224-225, 238, 241, 248, 251
  Affymetrix MAS algorithm, 81, 82, 83
Partial least squares, 33
PNNs, 151
Power, 50
Principal component analysis, 27, 71, 150, 210
Probabilistic neural networks, 151
Quantile normalization, 150
Robust multiarray averaging, 214
Sample size, 50, 125
Selection bias, 32, 173
Single value decomposition (SVD), 96, 99-100, 107-108, 148
Support vector machine (SVM), 148, 151- 153

Survival tree, 152, 155

MIX
Papier aus verantwortungsvollen Quellen
Paper from responsible sources
FSC® C105338

If you have any concerns about our products,
you can contact us on
ProductSafety@springernature.com

In case Publisher is established outside the EU,
the EU authorized representative is:
Springer Nature Customer Service Center GmbH
Europaplatz 3, 69115 Heidelberg, Germany

Printed by Libri Plureos GmbH
in Hamburg, Germany